2/7/91

AUDIO CONTROL HANDBOOK

AUDIO CONTROL HANDBOOK

For Radio and Television Broadcasting

Sixth Edition

ROBERT S. ORINGEL

Focal Press
Boston   London

To Andrew Shank, an engineer, someday . . .

Focal Press is an imprint of Butterworth Publishers.

Library of Congress Cataloging-in-Publication Data
Oringel, Robert S.
  Audio control handbook : for radio and television
broadcasting / by Robert S. Oringel—6th ed.
    p. cm.
  Includes index.
  ISBN 0-240-80015-X
  1. Electro-acoustics. 2. Sound—Recording and
reproducing. 3. Radio broadcasting. 4. Television
broadcasting. I. Title.
TK5981.07 1989
621.3841—dc19   88-20932

British Library Cataloguing in Publication Data
Oringel, Robert S.
  Audio control handbook : for radio and television
  broadcasting.—6th ed.
  1. Radio engineering
  I. Title
  621.3841
  ISBN 0-240-80015-X

Butterworth Publishers
80 Montvale Avenue
Stoneham, MA 02180

10 9 8 7 6 5 4 3 2 1

Printed in the United States of America

CONTENTS

PREFACE

This, the sixth edition of *Audio Control Handbook,* is a completely revised text. This edition follows the format of its predecessor editions, and it now has a new and much bigger publisher. We have moved to Butterworths' Focal Press, which also publishes our *Television Operations Handbook* and our coauthored *Access Manager's Handbook.*

We have, for this edition, augmented our knowledge of audio with that of Edwin Bukont, Jr., a very skillful practitioner in the audio field. Ed Bukont was commissioned to suggest changes, additions, and corrections to the fifth edition. His input to this volume has been extensive and very welcome. A graduate of Bowling Green State University, Bowling Green, OH, where, as a student, he used *Audio Control Handbook* in his classes, Ed has an excellent background for his assistance. He has been active in the recording, broadcast, theater, and concert fields and is a member of the Audio Engineering Society (AES). Ed currently is a special events and maintenance technician at the Voice of America, which is where I was employed, as both technician and supervisor, for 35 years. Ed additionally operates EB Audio, which does contract audio work. Ed Bukont's many and varied suggestions have made this sixth edition a much better book.

We have, as in past editions, relied extensively on printed descriptive literature and photographs supplied by the manufacturers of broadcast equipment products. We have quoted from or paraphrased that literature when necessary for clarity, with the permission of those manufacturers. *Audio Control Handbook* remains a "how-to" book rather than a technical treatise, however. It is still presented from the perspective of professionals in the field, and we hope that this approach contributes to the practical preparation and education of our youth for careers in the broadcasting industry.

We thank the manufacturers and their representatives and suppliers who lent their expertise and technical literature to this textbook. Particular thanks go to *Broadcast Engineering*'s "1988 Buyer's Guide and Spec Book," which provided current addresses for many of the equipment manufacturers. A word of praise, too, for Susan K. Blair of Datadog, Inc., Springfield, VA, is in order. Susan optically scanned the entire fifth edition of *Audio Control Handbook* and provided me with formatted computer disks of text files, making this revised edition easier to do.

The distinguished suppliers of equipment brochures, text, and photographs are as follows:

Murray A. Shields, vice-president of ADM Technology Inc.; Herbert M. Jaffe, vice-president of Atlas/Soundolier; Audio-Technica US; Lawrence J. Cervon, president of Broadcast Electronics Inc.; Cetec Vega; Lynn E. Distler, vice-president of Comrex Corp.; William A. Raventos of Crown International; David Moran and Lydia Capoccia of dbx Inc.; Laura J. Tyson of Denon America Inc.; Dolby Labs; Mike Dorrough of Dorrough Electronics; Ivan C. Schwartz of Electro-Voice; Arthur Constantine, vice-president of Fidelipac Corp.; Kelli Maag of Gentner Electronics Corp.; Rick Wanamaker of Gotham Audio Corp.; Hank Landsberg, owner of Henry Engineering; Tonnia B. Sills of HM Electronics, Inc.; William J. Parfitt of International Tapetronics/3M; Jack Kelly, president of Klark-Teknik Electronics, Inc.; Jim Cowan of Neutrik USA; Mr. Beuchat of Nagra/Kudelski, Lausanne; Lisa M. Vogl of Rupert Neve Inc.; Sidney R. Goldstein of Orban Associates Inc.; Sally Olson Saubolle of Otari Corp.; Peter F. Zarillo of L.J. Scully Mfg. Co.; Tony Tudisco, vice-president of Sennheiser Electronic Corp.; Tim Schneckloth of Shure Brothers Inc.; Mike Sakaguchi and Linda Metzdorf of Sony Corporation, Pro Products Division; Jean Kapen of Stanton Magnetics; Charles Conte of Studer Revox America; Herbert C. Klapp of Switchcraft Inc.; Jim Lucas of TEAC/TASCAM; Donald E. Mereen and Annette Peller of Telex Communications Inc.; XEdit Company; and Robert Trabue Davis of Yamaha Music Corp., USA.

All TASCAM brochure material and photographs are reprinted with permission of the TEAC Corporation of America.

1

INTRODUCTION TO AUDIO CONTROL

Audio control is the operation of all the various types of electronic equipment found in the studios, and the control rooms associated with those studios, of a radio or television station or a recording company facility.

The audio operator is the person who performs the control functions. This person is also known variously as the technician, the engineer, or the sound man. We refer to him in this text as the *operator,* to remain neutral in a controversy that has many work practice and salary ramifications that have nothing to do with the job. In addition, as much as possible we have tried to balance the use of masculine and feminine pronouns, knowing only too well that our industry abounds with extremely capable individuals of both sexes.

Most of the operator's work is performed in the control room, seated in front of and operating the audio control board or control console and its associated equipment, but occasionally she operates remotely, in the "field," with portable control equipment.

Sound, on entering a microphone, becomes transformed by the microphone into electrical energy. We refer then to the microphone as a transducer, a device that changes one form of energy into another.

Similarly, a loudspeaker, which changes electrical energy into mechanical energy (i.e., back into sound), is also a transducer. Occasionally throughout this text we may, for convenience, refer to sound as flowing through electronic components, although we are aware of this variance from scientific fact.

To begin, we will follow the path of an utterance from when it emanates from a performer's mouth with a microphone in close proximity to when it enters a listener's ears from a loudspeaker in close proximity. This is the audio path from the radio or television station to the home receiver (Figure 1–1).

As the performer speaks, sound waves are created and radiate in all directions from the sound source. The waves of sound impinge on many things, some of which absorb, some of which reflect, and some of which vibrate sympathetically with the sound. The sound wave has a pressure component (its strength) and a velocity component (its speed) expressed in Hertz (Hz) or cycles per second (cps). Sound waves in an elastic medium such as air follow the inverse square law: the intensity of the sound varies as the square of the distance. Thus, if the observer of the sound (the ear or a microphone) moves to twice the distance from the sound

1

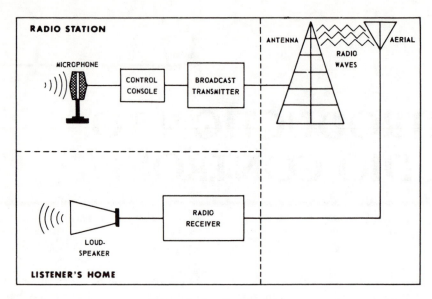

Figure 1–1   Radio path.

source, the sound intensity will decrease to one quarter of the original intensity.

The function of the transmitter in Figure 1–1 is to superimpose the audio energy onto a much higher frequency carrier wave in the radio frequency spectrum. This superimposition is called *modulation*. A carrier wave can be modulated in various ways, but the two ways most commonly employed in broadcasting are amplitude modulation (AM) and frequency modulation (FM). In AM the audio modulates or varies the size of the peaks and valleys of the space between succeeding waves of carrier. Figure 1–2 illustrates first an unmodulated radio wave, then an AM wave, and finally an FM wave.

Having modulated its carrier, the transmitter then feeds the combined energy to the antenna, which radiates it into the earth's atmosphere as electromagnetic waves.

The radio receiver also has an antenna. It experiences the transmitter's radiations, feeds them to an amplifier, separates the audio component from the carrier component, and then amplifies the audio en-

ergy to where it will, by transducer action, cause the loudspeaker to convert the electrical energy into sound waves.

TRANSITION FROM SOUND TO AUDIO SIGNAL

Sound waves created by a loudspeaker are analogs of the dimensions of the sound that created them. That is, the louder the sound source, the larger the wave. When a transducer changes those source waves to electronic signal, then the signal can be carried through the circuitry either in the analogous or analog form, where the size of the signal expands and contracts directly with the size of the sound. Alternately, the signal can be carried in digital form.

The introduction of the compact disk (CD) and its playback machines to home consumers, and consequently to radio and television stations, where the music is played in programming, focuses attention on this relatively new technology of digital audio. Actually, digital audio is more

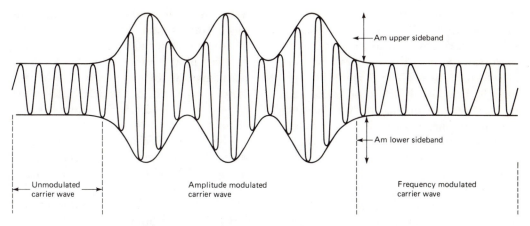

Figure 1–2   Unmodulated and modulated carrier wave.

than just one technology. It is a marriage of several technologies, among them audio, computer science (which includes mass storage of binary information), laser optics, error correction, channel coding, and digital filtering. More importantly, digital audio was conceived of and designed from scratch. It is not simply an advanced state of analog audio.

The attributes of digital audio that make it superior to its analog counterpart include its wide dynamic range, low distortion, ability to undergo limitless generations of dubbing and processing without significant degradation, and absence of interchannel time delay, or phase error.

This combination of technologies, which apply to music CDs and CD equipment, have similar application to rotary head digital audio tape (R-DAT). R-DAT, or DAT, is the upcoming wave of audio technology.

We present in this book an overview of the digital audio methodology, followed by a discussion of some of its types of equipment; we conclude with how digital equipment interfaces with the larger world of analog audio. A number of fine texts are available to expand the field of digital audio for the interested reader and present its intricacies in much greater detail. Highly recommended is *The Art of Digi-* tal Audio, by John Watkinson, which is published by Butterworths' Focal Press.

Let us begin our discussion by describing the terms *analog* and *digital* in an audio framework and then indicate how they differ in audio, keeping in mind that the concepts described are equally applicable to digital video or, for that matter, to any digitized signal.

ANALOG AUDIO

In analog audio the signal information may be varied in an almost infinite manner, within the boundaries of the 20 Hz to 20 KHz frequency band, and that variation may be described for our purposes in terms of voltage variation in audio circuitry or by magnetic flux strength variation in a magnetic tape recording.

An analog signal may be continuously amplified to reveal more and more of the signal's detail and complexity. The very process of amplifying an analog signal, however, inherently adds noise to the signal, until the law of diminishing returns applies and the signal is degraded beyond value.

For an analog signal to be true to the original sound, its conversion from sound to its electronic analogy must be linear. If the conversion is other than linear, harmonic distortion of the signal will occur. Further, during both recording and playback, the speed of the medium (the time at which it moves—33.3 rpm; 7.5 ips) must be held closely constant or the electronic/magnetic analog will not be true. An analog signal, then, is a continuous analogy of the original sound that is carried in two dimensions—time and amplitude.

In an analog system, all the signal degradations within the system are additive, and all are present at the output of the system.

In a typical analog audio system, consisting of a number of stages of audio, the finite limit on the number of stages is the amount of distortion and noise introduced into the signal within these stages, plus time delay instability of the signal such as is found in the group delay effects of inductors in the system. Thus, the more stages, the more amplification, but also the more noise and distortion, until the noise and distortion outweigh any possible value of amplification.

In an analog system, the noise and distortion introduced by the system itself can never really be totally divested from the original audio signal. Carefully designed and expensive filtering and processing equipment can partially remove noise and distortion, however.

Modulation

When analog audio is transmitted or stored, either its amplitude (AM) or its frequency (FM) is the determining factor in its modulation, or wave envelope.

DIGITAL AUDIO

Digital audio takes a completely different approach to the handling and storage of audio signal. Instead of a continuing analogy of the original sound, digital carries or stores a continuum of critically timed samples of the audio signal in a binary mathematical format, where one of two states—on or off, or alternately described as zero or one—define each sample's amplitude.

The two states change at specifically set times and are regulated by a stable timer or clock. Therefore, noise or time instability or both will be rejected at the point of reception, because the signal is retimed just before that point by the stable clock, which dictates all signal changes.

The two basic components of digital audio are sampling and quantization.

Sampling

As an analog signal, perhaps the output of a microphone, interfaces with and travels through a digital system, the original analog waveform is divided into evenly spaced time elements. This process is called sampling.

Each time element sample is thereafter expressed as an integer or whole number, which can be carried as a binary digit, usually called a *bit,* by computer people. A digital audio channel carries about 1 million bits per second.

In a digital audio circuit, the signal waveform is carried as though the signal voltage were measured at regular intervals with a digital voltmeter, or quantized, and the readings transcribed into binary math (i.e., zeroes and ones). The accuracy of the rate at which the readings are taken and the accuracy of the voltmeter completely determine the system quality, since once the signal voltage is expressed as a number, the number will be carried throughout the system without being changed. That is the key to the quality of digital audio.

In magnetic recording, the record or

playback head is oblivious to whether the signal is analog or digital, and therefore all the system's tape hiss, recorder/playback mechanical noise, distortion, signal dropout, print through, and crosstalk, all those negative aspects of analog audio discussed later in this text, are transmitted equally through the heads. In a digital system, however, only the variations in magnetic flux (the magnetic equivalent of signal voltage), represented by binary numbers, express the audio signal, and therefore all the noise and distortion are rejected.

In the transmission or storage of discrete audio samples, what happens between samples? Are we sure that when the samples are reassembled we will have a true reconstruction of the original audio waveform? Yes, if the original input audio signal was limited to a specific bandwidth. Sampling theory explains that a waveform which has been band limited and thereafter sampled can be accurately reproduced without knowing the intermediate values, if the amplitude values at the sampling points, and their positions in time are known.

Digital audio is not, however, without its own possibility for inherent error. To prevent two types of error specific to digital audio, called *aliases* and *images,* sampling must be done at a rate that is at least twice the highest audio frequency to be sampled, that is, more than twice 20 KHz.

The input signal to a digital audio system is filtered, so that no audio frequency above half the sampling frequency is admitted to the system. Pragmatically, the sampling of audio frequencies that are exactly half the sample frequency is avoided. CDs use a sampling frequency of 44.1 KHz, while the highest audio frequency admitted is 20 KHz.

When audio frequencies higher than half the sample rate are sampled (higher than 20 KHz), the sampling circuitry tends to become disoriented, generating aliases. An alias is a form of distortion that folds back into the audio spectrum of 20 Hz to 20 KHz. An alias that appears in the audio spectrum cannot be removed by filtering.

In addition to the audio frequency band of the sampled signal, the sampling process may create *images* of that frequency spectrum, outside and above the band at multiples of the frequency (i.e., 2f, 3f, and so forth) but since these images are well above the audio frequency band, they are easily filtered from the reconstructed output.

Errors may be caused in a digital audio system by tape dropout or by opaque material on the surface of an optical storage medium such as a CD. Major signal dropouts or heavy distortion may cause changes in magnetic flux or signal voltage, with a resultant error in the bits (numbers) recorded. This explains why a numerical error correction or concealment process must be used in digital systems. Since an error correction–concealment system is essential to return the numbers in the data stream to their correct value, digital audio would not be possible without it. Error-detection data are included in the digital bit stream, allowing the data stream's integrity to be checked at its destination, and errors are thereafter corrected or concealed. Since the data density of digital audio is so high, a minor tape dropout or speck of dust on an optical surface might cause the loss of thousands of data bits, so protection against error is essential.

As stated earlier, time is a critical factor in digital audio. Although digital signal as numbers can be easily carried by circuitry, mechanical devices such as recorders and playbacks may cause the numbers to appear at a fluctuating rate. Another essential for digital audio, then, is a temporary storage device that holds the numbers in the data stream and reads them out at a constant rate. The foregoing methodology, borrowed from television technology, is called *time base correction,* and its device, a *time base corrector* (TBC), is digital audio's stable clock. The time base corrector eliminates record or playback-head azimuth error, wow and flutter, and phase errors between recorded tracks caused by tape weave from a wobbly reel.

Finally, since a digital recording is no more than a readout of numbers, the recording can be dubbed through an infinite number of generations without any signal degradation between succeeding generations if the numbers are held constant.

Quantization

In practice, quantization is deemed to be less accurate than sampling because it involves approximation. The quantization process provides a specific number of discrete elements to represent the digitized signal amplitude. Any given sample must be represented by one of the available quantization levels. If none of the supplied levels precisely matches the true signal amplitude, then the system uses the closest available level. The difference between the actual signal amplitude and the quantization level used is called the *quantization error,* and although the error is usually very small, it can be audible, particularly on low-level music passages. It is sometimes termed *granulation noise.*

The number of quantization levels available depends on the digital word length, or the number of bits. The greater the digital word length, the greater the number of quantization steps, and concurrently the smaller the quantization error. For instance, the CD system uses a 16-bit word length that produces 65,536 quantization steps, leaving a very small margin of error between steps.

Since quantization error can be a source of noise in a digital system, the dynamic range of the system can be said to depend on the digital word length. The dynamic

range of a CD system is approximately 98 dB. Because we know how quantization noise is created, we also know how to prevent it. Strangely enough, the prevention method involves adding analog noise to the input signal. This added noise is termed *dither,* and it eliminates rather than masks quantization distortion. The amount of dither required is small but sufficient to lower the signal-to-noise ratio by 3 dB, which lowers the overall system dynamic range from 98 to 95 dB. Often, the background ambient noise in the arena where the microphone is located is enough to produce the dither needed.

Modulation

A form of encoding or modulation (the waveform envelope) is used to transmit or store digitized audio. The type of modulation most employed is pulse code modulation (PCM). PCM is superior to other modulation methods for digital audio because it depends on detecting the presence or absence of a fixed amplitude pulse or of a transition. In PCM the pulse, or transition, is either present or absent, enabling accurate decoding (demodulation) without considering noise in either the transmission or storage medium.

REVIEW QUESTIONS

1. Trace the audio path from the performer's mouth to the listener's ears.
2. What are the components of a sound wave?
3. What is a transducer?
4. Explain modulation of a carrier wave.
5. Explain the basic difference between analog audio and digital audio.
6. What is sampling, and at what frequency is it done?
7. What is quantization?
8. What is dither?

2

CONTROL BOARD EQUIPMENT BASICS

FUNCTIONS OF A CONTROL BOARD

The control board is the control center of audio operations. It will amplify, route, switch, pan, equalize, mute, measure, send, receive, and process audio signal. A well-conceived and -designed control board should be easy for the operator to learn and to use. Its controls should be ergonomic, and color coded for easy recognition.

A control board or control console has the following major functions:

1. It can amplify the relatively minute amounts of energy that it receives from the microphone outputs, record turntable outputs, tape playback outputs, and remote line sources, all of which are inputs to the console, to usable proportions. It can perform that amplification with an absolute minimum introduction of distortion or noise and minimal loss of frequency response.

2. It may remove distortion or noise or alter frequency response on input channels.

3. It can mix and balance the loudness proportions of the above-mentioned inputs, each against the others both singly and in subgroups.

4. It can control and route the amplified audio energy through program channels within the console to both internal and external points where program information is needed.

Control boards are electrically powered (or battery and alternating current [AC] powered, as in some portable equipment) and must be energized before use. The power switch for a console is often located on a nearby wall or other than directly on the console. In a control room new to him, an audio operator should ascertain the location of the power switch as well as the console's fuses or circuit breakers. Figure 2–1 shows the console as control center.

A BASIC ONE MICROPHONE–ONE PROGRAM CHANNEL CONSOLE

Described here are the basic features of a simple hypothetical console that amplifies and controls the output of one microphone (Figure 2–2).

Learn the names of the individual components of this console, because these components will be found repeatedly in the more complex consoles discussed in later chapters. Also, know the order of

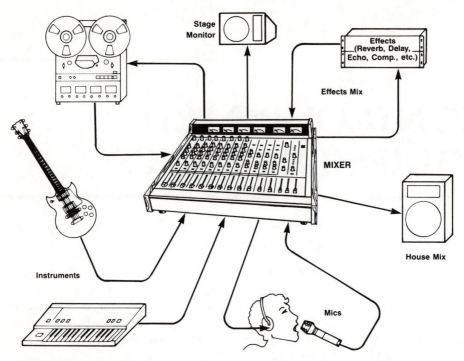

Figure 2–1   Console as control center. (Reprinted with permission from TEAC Corporation of America.)

the chain in which they are combined. Even the most complex control boards are extensions of this basic one microphone–one program channel console.*

Decibels and Volume Units

The control operator will refer continually to energy levels as she discusses audio equipment. Audio energy levels are commonly stated in volume units (VUs) and in decibels (dBs). The dB is one tenth of a Bel, named for Alexander Graham Bell. The Bel is too large a unit for ordinary use, so the dB is used. Decibels and VUs are of relative (not absolute) size. The dB expresses a ratio between two

voltages, two currents, or two power statements. Because it is a relative measurement, a reference level is always required. Thus the dB alone is referenced to 6 mW of power in a 500 ohm line. This is actually an obsolete reference base. The dBm, a more current measurement, and the one we really mean when we talk of a decibel, is referenced to 1 mW of power in a 600 ohm line that has a sine wave root-mean-square (RMS) value of 0.774 V across it. Both the dB and the dBm are ratios of sinusoidal power such as that emanating from an audio tone oscillator. Some other dB references are as follows: dBa, referenced above noise level; dBv, referenced above 1 V; dBw, referenced above 1 W; and dBx, referenced above crosstalk measurements. The formula for derivation of dB of power is dB = 10 log (to base 10) of P1/P2, where P1 and P2 are input and output power. When we talk of signal levels in a general sense, we talk

---

*Microphone is abbreviated in the audio industry as mic or mike. *Mike* will be used throughout this text.

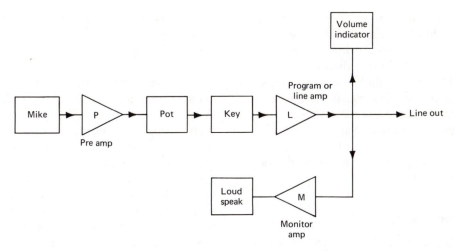

Figure 2–2   Block diagram of one mike–one program channel control board.

simply of dBs, saving the specifics of dBms and dBvs for specific reference.

The VU is stated as a change of one dBm for a complex waveform such as speech or music, as differentiated from a sinusoidal waveform.

In practice and in discussion, the operator uses the dB, the dBm, and the VU as relative absolutes in referring to audio power levels. For instance, the output energy of a mike is generally stated as being from −50 to −60 dB. The output of a preamplifier (known as a preamp in the trade) is generally from −20 to −10 dB, and the output of a program amplifier (or line amplifier) is usually +4 to +8 dB. The range of amplification or energy "gain" in the component chain varies from −60 to +8 dB, which is considerable, when one considers that a difference of 3 dB either halves or doubles audio power. Moreover, although the chain components are indicated as individual units, they are in fact electrically "hard wired" to each other in the order stated.

The mike's −50 dB energy output is fed to the preamp, which amplifies it to perhaps −10 dB. The potentiometer (known as a pot in the industry), on the output of the preamp, controls the preamp's output level or actual gain (and thus the mike's output level) from that high of −10 dB down to 0 in the same fashion that one might control the volume (gain) of a radio receiver.

The audio energy next feeds into the key (switch), which has an on-off, or insertion-muting, function, determining whether mike output feeds into the program amplifier. Closing or opening the pot with the key on provides the same function.

The program amplifier again enlarges the audio energy to +4 to +8 dB, sufficient to feed a transmission line to a program network or transmitter.

Across the output of the program amp is the VU meter, which enables the operator to visualize the console's volume output, and a monitor amplifier feeding a loudspeaker or earphones, which enables the operator to hear program content. Controlling program gain requires both watching and listening.

Diagramming Consoles

To discuss consoles, it is necessary to diagram their parts and functions using a combination of the block diagram and the functional diagram. The block diagram uses rectangular, square, oblong, or triangular blocks connected by lines with arrowheads to indicate direction of signal flow. Functional diagramming substitutes the schematic symbols used in electrical and electronic drawings for the less descriptive blocks. Figure 2–3 depicts some of these schematic symbols and their meanings. These standard descriptive devices are used internationally in the diagramming of circuitry.

Amplifier Theory

Next, let us *very lightly* discuss amplifiers. An amplifier enlarges an audio signal.

Amplifiers are the "active" components of electronic circuits and, for our purposes, the active parts of a control console.

Amplifiers can be described by either class or function. An amplifier's class—A, AB, B, or C—describes its electrical relationships and, more precisely, the portion of its input cycle, during which certain currents are permitted to flow under full load conditions. An amplifier's load is the circuitry into which that amplifier works. Amplifiers can be designed to amplify either signal voltage or signal power.

Describing amplifiers by function indicates what they do. Preamplifiers, which

Figure 2–3 Electronic schematic symbols. (1) Resistors, fixed and variable. (2) Capacitors, fixed and variable. (3) Coil. (4) Tapped coil. (5) Transformer, air wound. (6) Transformer, iron core. (7) Ground (earth). (8) Antenna. (9) Battery, generator. (10) (A) Earphones; (B) loudspeaker; (C) microphone. (11) Vacuum tube. (12) Transistor. (13) Amplifier. (14) Jack. (15) Key. (16) Switch. (17) Rotary switch. (18) Amplifier, patch connected. (19) Wires crossing, not connected. (20) Wires crossing, connected at dots. (21) Wires crossing, connected at A, not connected at B.

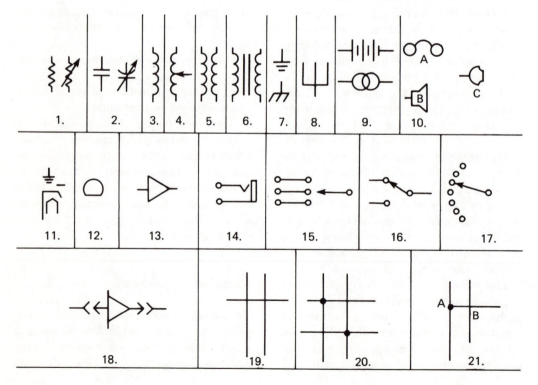

are voltage amplifiers whose design parameters allow for very high gain (+30 to +50 dB) and very low inherent noise and distortion, boosting the minute signal produced by a microphone to usable proportions. Power amplifiers, sometimes called current amplifiers, are designed to deliver many watts of audio power to a load.

Amplifiers are often verbally dissected for discussion by referring to their internal stages of amplification.

When we discuss how much work an amplifier can do, we at the same time discuss how cleanly it does that work by referring to its distortion level. The reader may by now have noticed the extreme preoccupation that professional audio people have with signal cleanliness. An amplifier's variance from absolute cleanliness of reproduction is stated in terms of a number of different factors. *Headroom* is the factor that describes the upper limit on gain that an amplifier can tolerate before it clips off the tops (upper and lower) of the waveform of a signal passing through. *Dynamic headroom* refers to the number of decibels above its normal power rating that an amplifier can tolerate on short duration bursts before it clips. This factor is measured at the amplifier's clipping level with a series of 20 msec tone bursts. *Clipping headroom* is a measurement of steady state operation, measured in decibels with sine wave signal fed to the amplifier. Some other measurements of an amplifier's abilities are its *slew factor*, which measures its transient response to sudden intense steep rises in frequency, and its *transient overload recovery time*, which measures its ability to settle back to normal operation after overdrive.

The most important measurement of an amplifier's ability to perform, however, is its *frequency response*, which describes the portion of the spectrum to be amplified versus the degree of uniformity over that spectrum portion that the amplifier displays in its coverage.

Last is an amplifier's signal-to-noise ratio (S/N).

A FOUR INPUT–ONE PROGRAM CHANNEL CONSOLE

Now we will expand our hypothetical one-microphone-input console to a more useful but still hypothetical instrument with four inputs to the program amplifier, sometimes called the program channel. These inputs, shown in Figure 2–4, include a second mike and two record playback turntables.

Note that the topmost chain of mike, preamp, pot, and key is identical to that of the initial basic console. We have now added three more such chains, all feeding a program buss, which in turn feeds the program or line amplifier. The buss is an electronically common tiepoint designed to remain as balanced circuitry whether any or all of the input chains are on or off, whether there is a load on it, or whether that load changes. Busses are used to "mix down" input channels from the many to the few.

On every console with more than two busses are assign switches that either send or block signal to the console's busses, depending on how the switches are set. Often an odd-numbered buss is paired with an even-numbered buss on the same switch. Then a panpot, or panning control in the input channel, determines how much of the audio signal goes to each of the odd-even combination. Busses are terminated at an output connection.

We have shown the key in Figure 2–4 functionally in each input. This enables us to introduce yet another element. In reality, the key can switch to the program buss, to the audition buss, or to off. Some consoles have more than one key or switch in the input, so that the input can be switched or assigned to a number of busses. The audition buss in Figure 2–4 feeds

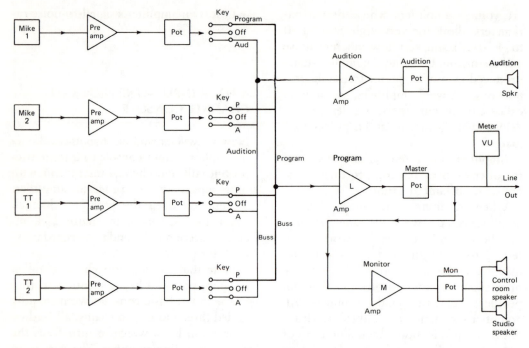

Figure 2–4   Block (functional) diagram of four input–one program channel console.

the audition amplifier and subsequently its audition loudspeaker.

Each input has its own pot, which varies its share of gain fed to the program buss.

On the output of the program or line amplifier we find a *master pot,* which controls the overall gain (volume) of the program channel. Across the output of the program channel are the VU meter and the monitor amplifier, this time expanded to include a monitor pot to control the volume output of two monitor loudspeakers—one in the control room and one in the studio. The monitor allows the operator to listen to the audio signal. The operator should be able to switch it to listen to any of the signals passing through the control board.

In sum, in this example of a console, the line or program amp feeds program signal to wherever it is needed, the audition amp permits previewing of signal be-

fore it is fed to a program source, and the monitor amp permits listening to the program source.

The console operator exerts control by manipulating the keys and pots on the console. The pot is the *gain controller* and is also referred to as a *mixer,* a *fader,* an *attenuator,* or a *variable pad.*

The operator uses the VU meter to observe and balance the levels of the various console inputs, to adjust the pots as she watches the meter, and to adjust the overall program channel output level.

Either or both mikes in this console may be used if the operator throws their keys to the on or program position and "rides gain" on the appropriate pot(s). Riding gain consists of turning the pot knobs (opening or closing them like water faucets) until the desired degree of volume is reached on the VU meter. The outputs of the record turntables are controlled in a similar manner.

The master pot on the output of the program amplifier must be open for the program channel to be used, and it can be used as a level controller for the entire program channel. Similarly, the monitor pot must be up to enable use of the aural monitoring facilities.

Finally, on some consoles, filtering and equalization components are present in the input chain and normally may be switched in and out of the chain as required. They are discussed in Chapter 3.

High-Level Inputs

Discussion to this point has centered around low-level inputs to the console (-50 to -60 dB). These inputs require a preamp to raise the gain to usable proportions. Console inputs deriving from the output of tape or disk playback machines, or from a remote line amplifier, will arrive at the console at perhaps a $+4$ dB level, with enough gain to drive the program amplifier. No preamp is needed in this sequence, so the high-level input chain will have a method of switching or routing around the preamp or will simply consist of input to pot, to key to program buss with no preamp. Some consoles have a switch in each input that permits a choice of high-level or low-level input device. Figure 2–5 shows a high-level input chain.

Potentiometers and Knobs versus Vertical Faders

The older, traditional large round knobs on console pots that open and close like water faucets have been mostly superseded by the more functional *sliding vertical faders* on many modern consoles, particularly those with the console face practically parallel to and rising up from the top of a desk.

These faders perform the same function as the traditional pots but appear on the console face as vertically marked rulers with a small sliding knob that can be moved up or down with one finger. Maximum gain on a vertical fader is at the top of the slide; minimum gain, or off, is at the bottom. Some vertical faders have a detent position, one click past bottom, that places the input chain in cue position.

Vertical faders are advantageous on a console with many inputs because the operator can tell at a glance which pots or faders are open and to what degree relative to each other. Vertical faders enable the operator to react more quickly during a rapidly paced program. Buss faders are often color coded differently from input channel faders for differentiation.

Audio energy flows from a console's input channels to its busses (Figure 2–6).

Audition, Cuing, and Talkback Systems

When a microphone or a turntable input key is in the *on* or program position, that chain is feeding *live* into the program channel and perhaps on the air.

To enable an announcer or other talent in the studio to talk to the control room through a studio mike without that mike feeding the program channel, and while a tape or record is feeding the program channel, a third or opposite position, the cue or audition position, is provided on the microphone keys.

Figure 2–5 Block diagram of a high-level input chain.

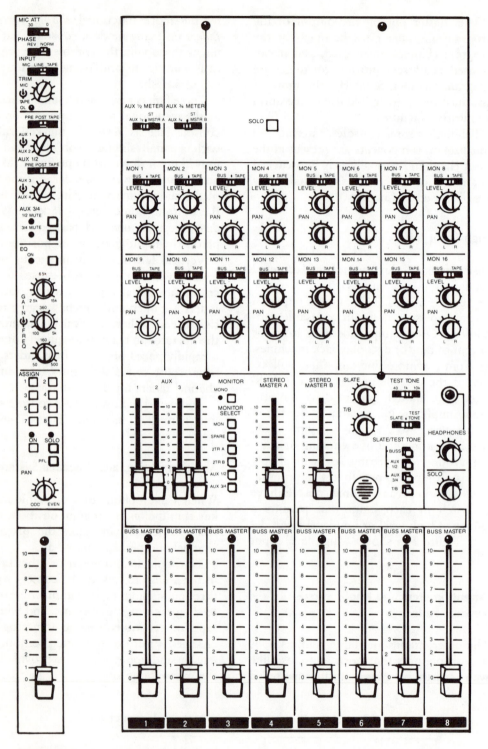

Figure 2–6 Input channel strip and busses. (Reprinted with permission from TEAC Corporation of America.)

Each input key on our hypothetical console has three positions: program, off, and audition. Input keys are mounted on the front face of the console, each above its corresponding pot, and they operate— are thrown—either vertically or horizontally depending on console design.

Consoles with vertical faders have push button switches instead of keys. Some have a detent switch for cuing at the bottom of the fader. The operative push button is generally illuminated when activated to indicate its function. Those consoles with a set of push button assign switches allow the operator to assign an input to a program or an audition buss or to a subgrouping of inputs. Stereo input consoles may have a panpot control, which permits panning of the input between two (left and right) stereo channels.

The audition position of the key on turntable or tape inputs permits the operator to audition or listen to this input without placing it on the program channel. The same audition position is used for cuing a record or tape, which is discussed in Chapter 10.

Throwing the input key to the cue position feeds that input to another amplifier in the console known as the *cue* or *audition amp*. This amplifier feeds a separate loudspeaker called the *cue speaker*. The operator listens simultaneously to the program channel on the monitor speaker and to any other input that he may wish to audition or cue on the cue speaker. A separate VU meter is often available across the audition amplifier for visualizing levels while cuing.

Since the announcer in the studio is able to converse with the control room (provided that the operator throws the mike key to the audition position), then the audio control operator must be able to respond. She performs this function through a *talkback* mike in the control room, operated by a spring-loaded key or push button on the console. The key or button

springs back to its off position when released.

A note of caution here: If one mike in the studio is live, that is, feeding the program channel, then no other mike in that studio, or the control room talkback mike, may be used on the audition system. Clearly, anything said into the audition mike would also be picked up by the live mike.

Each studio incorporates a monitor speaker as part of its equipment to enable its program participants to hear recorded portions of the program. Whenever a mike is live in that studio, however, the monitor speaker is disabled automatically by a muting relay in the console, energized by yet another function of the microphone key. If the speaker were left on with a live mike in the room, a feedback loop between the mike and the speaker would instantly occur. Feedback, which is a loud oscillatory howl, occurs through repeated amplification and reamplification of a signal or sound going through microphone to console to speaker back to microphone. Feedback can be overcome rapidly by either turning the studio monitor pot off or by cutting the microphone key. The latter action, of course, cuts off program as well, so it is preferable to turn off the monitor until the faulty muting relay can be cleared. The British call feedback *howlround*.

The VU Meter

The VU meter enables the operator in the United States to monitor sound or sound intensity by measuring the average value of that intensity. This permits him to balance the intensities of two or more sound sources by varying their pot settings on the control board. The meter is additionally used to control the overall volume output of the program channel, using the master pot. The VU meter has no function in measuring the frequency response of the sound inputs.

On a single program channel console, the VU meter is generally located centrally on the upper face of the control board. A stereo console has two VU meters, left and right channels, side by side.

The standard VU meter has specially designed damped ballistic response to enable it to respond to average sound volume rather than to individual sound peaks. The meter face has a dual scale. The upper scale is read in volume units, from −20 to 0 and then from 0 to +3, in red. The 0 is the maximum normal reading without distortion. Beyond 0, the scale is marked in red to indicate that readings in this zone are "hot." Readings of volume in this area should only be permitted momentarily. We say that we are "in the red" or "spilling over" when the meter reads in this area. When the meter averages −20, we say we are operating "in the mud."

The lower scale of the VU meter is marked from left to right and from 0 to 100 at full scale and reads percentage of modulation.

The operator then uses the VU meter to ride gain on the program channel. Riding

gain consists of watching the meter, listening to the monitor speaker, and adjusting the pots as necessary to maintain appropriate audio levels. Just what levels are considered appropriate will be discussed in Chapter 9. However, since the standard VU meter virtually ignores peak readings, operators compensate by riding dialogue from 3 to 5 dB below music to control peak levels. Another practice is to allow for a *crest factor* of at least 10 dB to cover the peak excursions of the audio waveform, which the VU meter is too slow to indicate. Figure 2–7 shows a VU meter.

Other Visual Level Measuring Devices

Although the VU meter, with the needle reading across a scale, is the visual sound measuring device most often used on consoles made in, or used in, the United States, other visual meters are available.

One such device is a light-emitting diode (LED) horizontal bar graph. This meter lights up LEDs on a line from left to right, corresponding to audio level intensity, from −20 to 0 in yellow or green LEDs and from 0 to +3 VU in red LEDs, with the same average reading characteristics as a VU meter. The LED meter can also be designed to read vertically or to have peak program meter (PPM) characteristics.

The PPM is used in the United Kingdom and some European countries. It has a needle indicator like the VU, but its ballistics are such that it reads audio level peaks rather than averages. British audio operators generally subscribe to the view that VU meters underread peak levels, typically by 10 to 12 dB, and thus introduce distortion. Some newer VU meters have incorporated an LED peaker, built into the meter's upper right corner, that flashes red on overload peaks. One manufacturer builds an LED bar graph into

Figure 2–7   Volume unit meter.

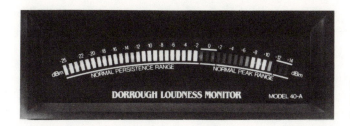

Figure 2–8 The Dorrough loudness monitor. (Courtesy Dorrough Electronics.)

and above the VU meter face. Where a multiplicity of different audio levels must be viewed at one time, they may be displayed on vertical bar graphs displayed in real time on a color television monitor.

The Dorrough loudness monitor (Figure 2–8) displays the actual energy content of program material by integrating the peak and average amplitudes of a complex waveform program. In addition, it can integrate both channels of a stereo signal.

Figure 2–9 indicates how the basic four input–single program channel console might look externally if constructed.

Submixing and Subgrouping

As pictured, the console has minimal operating features. To operate effectively, the audio operator will need more than the two mike inputs. He will also need, in addition to the two turntables, inputs for open reel tape, cartridge tape, cassette tape, incoming remote lines, and network lines. He will need a minimum of 10 to 15 inputs at a small radio station or in a single television control room and up to 50 inputs at a complex recording facility. Some of the inputs may be switchable to one of several types. Further, if his input needs are complex, he may need facilities for subgrouping the inputs within the console. A submix facility can accept assigned mike input chains and group them on a submix buss. The operator may also need signal processing facilities in the console's inputs. Figure 2–10 illustrates a five-mike input submixer.

The submix buss feeds through a console submix master pot that feeds a console program buss. The console's metering, auditioning, and monitoring facilities handle those functions for the submixer.

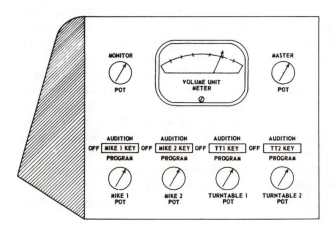

Figure 2–9 Basic four input–single program channel console.

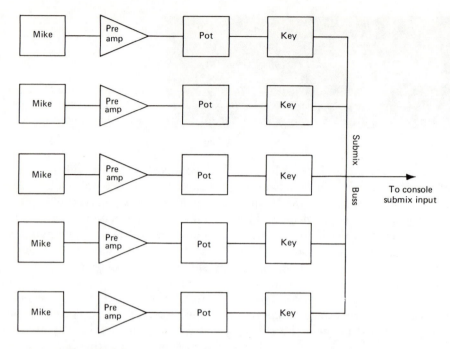

Figure 2–10 Block diagram of a five-mike input submixer.

MULTIPROGRAM CHANNEL CONSOLES

We have attempted to describe only single program channel consoles, but there are, of course, consoles with multiple program outputs. A single program output is satisfactory for AM radio or monaural recording, but two program output channels are necessary for FM stereo radio, for stereo recording, or for using a single console control room to handle two studios simultaneously.

A multitrack recording facility requires a program output channel available on the console for each track to be recorded or to be played back during mixdown, which is often as many as 72 program outputs.

A multiprogram channel console has all the features of a single channel board, but the features of the input chain will either be duplicated for or switchable to each

program channel. For instance, an eight-program channel console might have eight VU meters; or two VU meters, each switchable to one of four program channels; or two VU meters switchable to either audition channels or program channels. A modular multiprogram channel console will have, as part of each input module, a method of switching the output of that module (assigning) to a subgrouping facility or directly to any or all of the program or audition channels.

Additionally, there may be inputs that are switchable to either line level or mike level, panpots that fade an input from one channel to another, and effects (e.g., reverb) to add spatial dimension to music or speech through time delay or echo. There may be preview, or solo, which enables the operator to listen to a single isolated input separately, without taking it out of the mix that is feeding the program

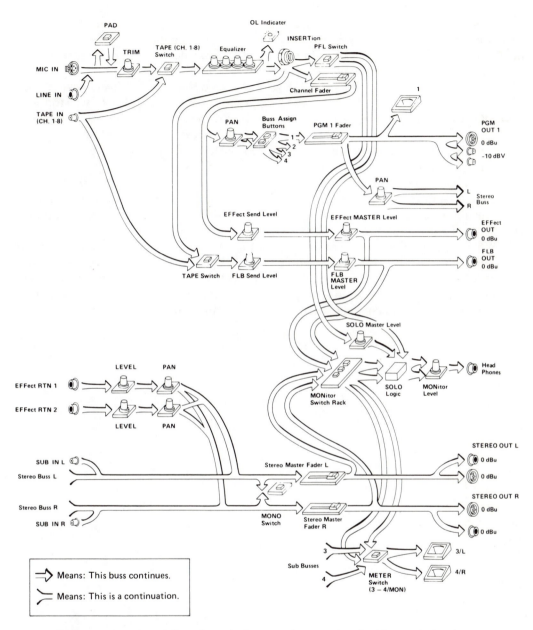

Figure 2–11   Signal flow through the console. (Reprinted with permission from TEAC Corporation of America.)

channel, and foldback, a feed to the on-stage performers for their music-balancing requirements.

Figure 2–11 shows how audio signal flows through a particular control board.

The next step in our discussion of consoles is to examine some representative professionally manufactured consoles, but before we do that, note that adding dimension to, or processing, sound is one of the functions of input modules. We will therefore look at audio processing equipment in Chapter 3 before we explore the larger picture of professional consoles.

REVIEW QUESTIONS

1. List the four major functions of a control board.
2. List in normal sequence the units in a mike input chain and describe the function of each.
3. What is a dB, a dBm, and a VU?
4. What are four other names for pots? What function does the master pot have?
5. What is an amplifier's load? Its headroom? Its frequency response?
6. What is a program buss?
7. Define riding gain.
8. What do we use a cue system for? A talkback system?
9. Describe audio feedback. What causes it? How should we stop it?
10. What do we use the VU meter for? What do we not use it for?
11. Describe some other level measuring devices.
12. What is a vertical fader? What advantage does it have over a knob?
13. What is submixer? Input subgrouping?
14. What is a multiprogram channel board? In what way does it differ from a single program channel device? Describe some uses for a multiprogram channel board.

3

AUDIO SIGNAL PROCESSING

Audio signal processing, perhaps borrowing its name from electronic data processing, and indeed using many of the same types of systems and devices, has made major inroads into modern audio control equipment. This text will broadly define audio processing equipment and technique as the use of a number of diverse electronic devices to alter the audio signal to remove unwanted portions of the audio spectrum, to remove noise or distortion, and, either deliberately or incidentally, to add gain or boost to specific portions of the spectrum. Note that we use terms such as *alter* and *remove,* but processing never adds anything except gain. The audio spectrum for our purposes is 20 Hz to 20,000 Hz (20 KHz).

A relationship exists between an amplifier's output power and its degree of distortion, that is, how much work an amplifier can do versus how cleanly it can do it.

Among the devices used for analog audio signal processing are audio equalizers (EQs, pronounced like the letters), low-frequency extenders, companders (compressor-expanders), noise reduction systems (dbx, Dolby), noise gate circuitry, and time-delay devices. Among digital processing devices is a multieffect processor by Yamaha.

EFFECTS AND AUX SYSTEMS

Some processing devices are found within the console, and some are external stand-alone devices. The external devices, often found in racks in the control room, require a way to receive console audio signal and a way to return processed audio to the console's signal path. Processing is often called "effects" for the effect it has on the audio, and AUX systems (auxilliary inputs, auxilliary busses, and auxilliary outputs) are what move unprocessed audio out and processed audio back in again.

Earlier in broadcasting and recording, equalizers were used mostly as stand-alone devices, to be patched into existing audio control equipment, and they were used solely to eliminate signal distortion or noise frequencies. This unwanted noise or distortion may have been picked up or self-generated anywhere in the audio signal chain. Another common use of audio equalizers was to alter the acoustics (reverberation time) of the studio or auditorium where the program originated. Audio equalizers are still used for these purposes, but they have become a programming tool to alter the program source's basic sound. In the mixdown session following a stereo music recording, audio equalizers can be used to correct what the recording director or producer sees as musical deficiencies. This process, called *sweetening,* moves the director (and the control operator, who is his "hands") from observer-translator-recordist to someone who affects how the program material sounds.

Needless to say, this development has spawned two highly divergent and vocal schools of thought on how much engineering should be part of a recorded music selection. Certainly there is no disagreement that rock music is as much, if not more, the product of electronic devices than of traditional acoustic music instruments.

Most modern control consoles have built-in equalizers that can switch into and out of virtually every input channel. This permits the console operator a wide variation of control over the quality and musicality of the sound entering the program mix. Although this control feature is desirable and versatile, remember that performing a complex mix, controlling only the audio gain without equalization in every input, is itself difficult. Adding to this mix the profusion of equalization in differing combinations in each input places the operator in jeopardy of getting lost in a maze of equalized close-miked sound. An admonition to keep it simple is in order here. An operator in the throes of a fast-moving, rapidly changing audio scene can only keep in mind just so much information with which to make fast console control decisions. If she is controlling a large number of microphones acoustically coupled tightly to an equal number of musical instruments or orchestral sections, each equalized to the hilt to get "just the right sound," then if the director decides that the sound is not quite what he wants, the operator must quickly ask herself, "What sound did he like, and how can I duplicate it?" Since human ears have no memory, the answer may be difficult. The dilemma may be solved with a con-

sole that has microprocessor-assisted memory. Microprocessor assist is a built-in feature of some large consoles with 30 or more inputs. Other consoles are designed for computer interfacing.

To understand the process used, the reader should understand that making music recordings is as much a business process as a musical or operational one.

Let us first look at the monaural music recording where the operator-recordist has but one shot at a perfect microphone pickup for each performance that is recorded. In fact, two performances of the same piece are usually recorded under identical (or so one hopes) conditions so that minor musical errors and audience coughs during low passages can be edited out before the recording is released to the public. The orchestral balance, mike placement, and mike levels must be perfect each time because nothing can be changed after a public performance. If the performance takes place in a recording studio, any changes made in mike placement or console orchestral balance are very costly.

To save time and money for multitrack stereo recording, a wholly new approach was conceived. The music group, be it a four-piece combo or a full symphony orchestra, is divided for miking purposes into separate entities—sometimes individual instruments, sometimes sections or groups of instruments, with each entity picked up by a separate mike. Then that mike is fed into one track of a multitrack tape recorder. All tracks on this recorder are recorded at about the same audio level. If the orchestra were divided into 24 entities, then 24 mikes would be used to feed 24 tape tracks. What the operator-recordist gets out of the recording session or performance is most certainly not a finished product but a 24-track recording of raw material that will later be mixed down, equalized, balanced, and shaped during a session or sessions called the *mixdown.* The mixdown of the 24 recorded tracks will be done through a control console, and those who operate and direct it will end up with a piece of recorded music that may resemble what the orchestra played or may not.

The 24 tracks of taped music are fed to 24 separate console inputs and this time are mixed down, perhaps a pair at a time, to 12 console program channels feeding 12 tracks of a tape recorder. Then the 12 become six, and what finally emerges are two tracks of stereo music. The more tracks employed initially, the more flexibility in mixdown and the more music decisions to be made. The operator or the director should have a musician's assistance during this process.

In this recording context, the inputs to the console all have discrete equalizers so that each input track of taped raw music material can be shaped and controlled.

In mixdown the operator plays the original 24-track flat level recording through the console, first making obvious balancing moves (e.g., violins are not as loud as drums), then balancing section with section and adding differing degrees of equalization at the director's instruction, and at the same time rerecording the mix. If, in mid-piece, the director is unhappy with the recording mix, the mix must be rebalanced and rerecorded. As this process is repeated a number of times, with different level and equalizer settings, and as the operator is asked to go back to a previous starting point with still different level and equalizer settings, computer assist becomes necessary.

The computer-assisted console is designed so that whenever the operator adjusts the level or equalizer controls, the computer stores the settings in its memory and can duplicate them on demand by servo control pot movement.

Note that there are some distinct disadvantages in analog multitrack recording. First, there is 3 dB of added noise per track, per tape, regardless of the noise-reduction format used. Thus, a 24-track tape contains 72 dB more noise than a single track. When mixed down to six tracks, 18 dB more noise is added to the signal. The message here is that just because a recording facility has a 16-track recorder, you need not use all 16 tracks to get your money's worth.

Another negative consideration is that when recording stereo miked music on multitrack recorders, a phasing error may be introduced into the stereo group by a poorly wired studio, a poorly aligned tape recorder, or, more often, by the natural phasing of the mike pickup. Serious music recordists should require phase-reversal switches on console inputs and mix-minus provision on the console, and they should listen to monophonic mixes of a stereo program. Special attention should be given to the left plus right (L + R—the sum) and left minus right (L − R—the difference) between any two channels. Sum represents the monophonic version of the stereo tracks. This is what listeners in monophonic AM radio and television will hear. When listening to the sum, what is heard is all the audio that is common to the left and right channels. If the mix is intentionally for mono, there should be no difference between the sum monitor and what is heard from just the left or the right monitor. The difference monitor contains all audio that is not common to the two channels. In an intentionally mono mix, this monitor should be silent. If more than a residual amount of audio is heard in this monitor, it may indicate phasing error. In a stereo mix, this difference monitor will contain reverberation and solo program (material in only the left or only the right channel), but, as in mono, a significant amount of the primary program material should not be present.

AUDIO EQUALIZATION AND FILTERING

An audio equalizer is a combination of audio filters in one stand-alone device or on one module type in the console. It can be either passive, containing no amplifiers, and thus present its insertive circuit with a dBm loss, or it can be active, which means that it includes amplification. The added amplification can be of unity gain, sufficient only to replace the gain lost in the equalizing process, or it can incorporate additional gain, provided specifically to frequencies that it wishes to boost.

Filters are composed of reactance networks, which are designed to perform low-pass, high-pass, low-cut, or high-cut functions in terms of audio frequency spectrum portions. Reactance, either inductive or capacitive, combines with resistance in an alternating current circuit to create impedance. The term *pass* means that the frequency band in question is allowed through the filter, while the obverse, *cut,* means that the frequencies in question are not permitted through the filter. Functionally, filters either attenuate or boost the gain of specific sound frequencies, usually in three-decibel steps at a particular rate or slope of the attenuation curve, which is stated in decibels per octave. Some typical curves are presented later. The slope is determined by the order of the filter (first order, second order), that is, how many capacitive or inductive reactances comprise the filter. The higher the filter's order, the steeper its curve's slope and the more complex the curve's attenuation or boost.

When a filter is designed to pass a specific band of frequencies, cutting out all others, it is called a bandpass filter, and that band of frequencies is called the passband. The width of the passband, or portion of the audio spectrum that is passed, is determined by the Q number of the filter. The Q can be termed the degree of narrowness of the band. The higher the Q number, the narrower the bandpass.

The obverse of the bandpass filter is the notch filter, which passes all frequencies except for a very narrow passband consisting perhaps of only one specific frequency. Notch filters for approximately 60 or 120 Hz, the typical United States electrical line hum, which tends to creep into audio circuits, are very common.

A discrete audio equalizer, then, is a device comprising a number of filters packaged in a unit whose front panel layout is designed so that the operator can adjust the gain or slope of a series of those filters ranging across the audio spectrum. If the controls are arranged in one-third-octave steps with slider pots graphically arranged from a center position, it is termed a *graphic equalizer*. This type can equalize a large number of frequency ranges (27 on the Klark-Teknik) at the same time.

The two discrete types of equalization are *shelving* and *peak-dip*. A shelving equalization curve changes in level until it reaches a specific frequency; then it flattens out and does not change again. The flat portion, which looks like a shelf, gives the curve its name. The peak-dip equalization curve looks like a mountain or valley. It starts at zero and climbs to a peak as it boosts frequency gain and drops to a sharp dip or valley as it cuts it. Graphic equalizers usually peak-dip, with the controls changing gain at specific frequencies and the Q fixed.

A graphic equalizer's center positions usually follow standards set by the International Standards Organization. For a one-third-octave equalizer they are 16, 20, 25, 31.5, 40, 50, 63, 80, 100, 125, 160, 200, 250, 315, 400, 500, 630, 800, 1000, 1250, 1600, 2000, 2500, 3150, 4000, 5000, 6300, 8000, 10,000, 12,500, and 16,000 Hz.

Klark-Teknik Graphic Equalizer

The Klark-Teknik DN 300 graphic equalizer, made in the United Kingdom, is a one-third-octave, 30-band device providing boost or cut of up to 12 dB at 30

Figure 3–1   The Klark-Teknik graphic equalizer. (Courtesy Klark-Teknik Electronics, Inc.)

different frequencies covering the entire audio spectrum. The vertical fader controls have 0 dB center detents or stops. An equalization bypass switch on the right side of the rack-mounted unit connects input directly to equalizer output. The gain control, above the bypass switch at the extreme right, sets input level from infinite attenuation to +6 dB of gain, with a unity gain center position. A ground lift switch enables signal and chassis grounds to be isolated, eliminating ground loop problems. To the left of the gain control are adjustable high- and low-cut 12 dB per octave shelving filters, with a selectable 6/12 dB per octave high-cut slope. A security cover is available for permanent control settings where system calibration has taken place. Figure 3–1 shows the Klark-Teknik DN 300 graphic equalizer.

Orban Equalizer

Another style of equalizer, the Orban 622 (Figure 3–2) is a parametric equalizer. It provides continuously variable control over the three fundamental parameters of equalization: the amount of peak or dip in decibels; the center frequency at which the maximum peak or dip occurs; and the

bandwidth, or number of frequencies on either side of the center frequency affected by the equalization. In a true parametric equalizer, adjusting a single parameter does not affect the other two parameters. The 622 is a two-channel equalizer with four noninteracting peak boost-cut sections, each with bandwidth and tuning control. The equalization range is from +16 to −infinity dB. The Q is adjustable from 0.29 to 3.2. There is 12 dB of gain available on front panel controls, in-out switches for each section and for the entire equalizer, and an overload LED indicator for each channel.

The curves in Figure 3–3 indicate the boost or attenuation available as equalization is accomplished by summing the output of a bandpass filter to the main signal in phase for boost or out of phase for cut. As the bandwidth control is operated, the skirts of the equalization curve move in and out, but the peak gain remains constant. As the tuning control is operated, the curves slide along the frequency axis, but their shape is unchanged.

Pink Noise

A most dramatic use for audio signal processing is in the transformation of an

Figure 3–2   The Orban 622 parametric equalizer. (Reprinted with permission from Orban Associates, Inc.)

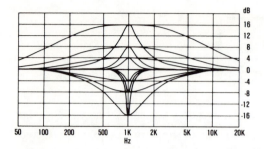

Figure 3–3   Orban 622 equalizer curves. (Reprinted with permission from Orban Associates, Inc.)

ordinary subscriber service telephone line. The extensive use of beeper telephone feeds to radio and television stations for relatively inexpensive line coverage of special-events programming (e.g., an out-of-town basketball game), which are used to save the station the costs of program-quality lines, and the equally extensive use of telephone feeds on listener call-in programs, both present the station with initially low-quality, noisy, distorted audio in the 200 to 2000 Hz range. Judicious use of signal-processing devices can virtually eliminate the noise, clear the distortion, and make the incoming beeper signal appear to sound cleaner. Please note that the operative word is *appear*. Signal processing does not add frequencies that were not there. What it does is boost the gain of the desired low-gain frequencies, eliminate unwanted frequencies, and rebalance what is left after the noise and distortion have been filtered out. The term *beeper* derives from the days when the Federal Communications Commission (FCC) required that a beep tone be transmitted on a telephone line being recorded so that the party at the other end of a conversation knew for certain that his conversation was being recorded or broadcast. The requirement no longer applies and beeps are not transmitted, but the name remains.

To judge the difference between an unprocessed telephone line circuit and one that has undergone processing, the operator may use, in addition to her ears, a spectrum analyzer, which gives a readout either in an LED or on a cathode ray tube (CRT) display. In each case, the operator can view the telephone line or, more precisely, an audio signal on the line both before and after adding signal processing devices to the line. The test signal that she examines is called pink noise.

All the frequencies perceivable to the human ear, when heard together, are called white noise. If a generator of white noise signal is fed into a telephone line, with a rising characteristic of 3 dB per octave, through a special audio filter that has the inverse frequency characteristics to that of white noise, what results is a signal with uniform level across the audio spectrum called pink noise. This signal, which is discernible to the ear as a high-frequency hiss, is used solely for measurement. The uniform level of the pink noise fed to the line is reshaped from a flat curve by any line limitations. The spectrum analyzer first reads the pink noise's flat curve and then reads the reshaped signal, displaying its variations as bends or excursions of the flat curve. The Klark-Teknik spectrum analyzer shown in Figure 3–4 displays 30 frequencies over the normal human hearing curve from 25 Hz to 20 KHz, using LED columns for the display. The frequencies chosen match the control frequencies of the Klark-Teknik 300 series equalizers. When a signal overloads a display column, the two lowest LEDs go dark. Relative levels are shown by the two scales to the left of the display. To the right of the display, the memory button selects one of three memories, each allowing a complete analysis to be stored and displayed at any time. Three LEDs show which memory is currently being read. The average-peak switch (AVG-PK) changes the ballistics of the display and selects

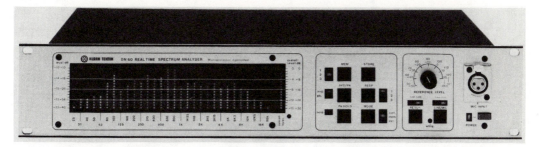

Figure 3–4 The Klark-Teknik DN 60 spectrum analyzer. (Courtesy Klark-Teknik Electronics, Inc.)

either an averaging or a quasi-peak characteristic. Peak hold (Pk Hold) activates a separate LED on each column, which memorizes the peaks. The continuous real time display is still valid. Pressing store updates the peak display. The store switch writes current input to the selected memory, clearing the previous content. Response time switch (RESP) enables a wide range of signals. Mode switch (MODE) selects either real time operation or one of the three memories. The reference level control sets base level for either display in dBm or for mike input in dBspl. The XLR-3 at the extreme right is for the Klark-Teknik AT1 measurement mike. On the rear panel of the DN60 are two XLR-3s: a line input connector for any line level signal and a pink noise output connector. If it is required or desirable, a signal-processing device such as a graphic or parametric equalizer is then placed in the line to reflatten the curve.

Gentner Engineering Extended Frequency Transceiver

Gentner Engineering's EFT 900 is a low-frequency extender system for telephone lines that shifts its entire input signal upward by 250 Hz in the transmit or encoder end and downward by 250 Hz in the receive or decoder end. Thus, 100 Hz becomes 350 Hz (100 + 250) in the telephone company line portion of the circuit, and 2000 Hz becomes 2250 Hz. This shift takes place across the entire transmitted portion of the audio spectrum and improves telephone line audio quality by recovering 2.5 octaves of low-end frequency response. Since most of the noise in telephone circuits is at the lowest frequency end, the circuit seems to be improved by 250 Hz precisely in the most noisy area.

The EFT 900 includes an internal coupler to the telephone line, a built-in mike preamp from its mike input, a line input, and an internal headset amplifier. Figures 3–5 and 3–6 illustrate the Gentner EFT 900.

dbx Noise-Reduction System

Before we describe the equipment used in the dbx noise-reduction system, we will discuss noise-reduction theory.

There is a continuing problem of maintaining balanced audio quality when switching from a live mike to records, cartridges, or remote coverage of live events. The live mike sounds good, the vinyl records are good except for scratch and surface noise, the high-frequency end on the carts becomes distorted with saturation, and the remote feeds contain hum, hash, and other noise. A noise-reduction system of the type made by dbx,

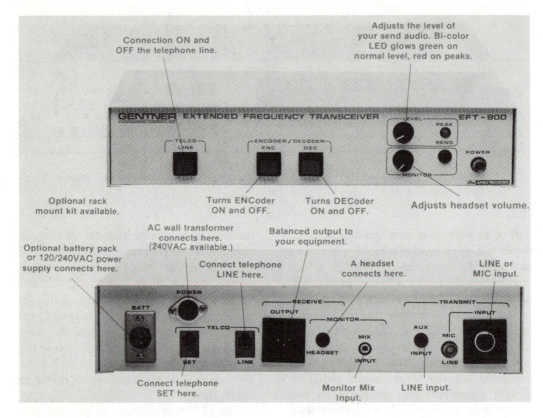

Connection ON and OFF the telephone line.

Adjusts the level of your send audio. Bi-color LED glows green on normal level, red on peaks.

Optional rack mount kit available.

Turns ENCoder ON and OFF.

Turns DECoder ON and OFF.

Adjusts headset volume.

AC wall transformer connects here. (240VAC available.)

Balanced output to your equipment.

Optional battery pack or 120/240VAC power supply connects here.

Connect telephone LINE here.

A headset connects here.

LINE or MIC input.

Connect telephone SET here.

Monitor Mix Input.

LINE input.

Figure 3–5 The Gentner EFT-900. Front *(top)* and back *(bottom)* views. (Courtesy Gentner Electronics Corporation.)

called a compander, is a combination of compressor and expander, which is essentially what the system does. It compresses the program signal before it reaches the noisy media described above, and it ex-pands that signal after it departs the noisy media. Noise reduction takes place because of the masking effect of the compressed, and therefore higher level, program signal on the noise level. With-

Figure 3–6 The Gentner EFT 900, block diagram. (Courtesy Gentner Electronics Corporation.)

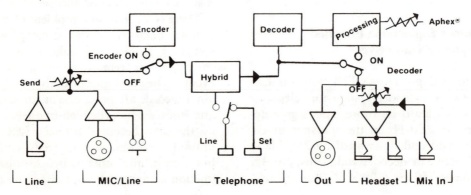

out noise reduction, high-level music signal will normally be loud enough to mask one's perception of noise. During quiet music passages, however, noise becomes audible. Placing the compressor before the source of noise keeps the signal going through the noisy medium at high signal level so that masking can occur. The expander at the other end then restores the signal to its proper value, pushing the noise down at the same time. During periods of silence, the noise will be pushed down far enough to be virtually inaudible. The noise-reduction system can also reduce distortion, because for a given signal-to-noise ratio, audible levels can be reduced compared with a similar situation without noise reduction. Lower signal levels almost invariably result in less distortion of the signal. The normal degradation of the signal-to-noise ratio with less signal will not take place because of the noise-reduction system.

This simplified explanation omits the sophistication required in such a system to cope with signal time delay, where the relative phase shift of different frequencies may markedly alter signal wave shapes.

Let us look now at the dbx noise-reduction systems. Type I is for professional-quality audio tape recorders, and type II is for nonlinear frequency response tape decks, cart machines, video cassette recorders (VCRs), and consumer-grade equipment. The two systems are incompatible, but each type doubles the dynamic range of the transmission medium to more than 115 dB. Each unit therefore

can reduce the noise of the medium by 40 dB or more.

dbx noise reduction consists of an encoder that compresses its input signal at a constant 2:1 compression ratio and applies a carefully tailored frequency response preemphasis during record, followed by a complementary expanding or decoding of the signal 1:2 with a precisely complementary deemphasis during playback. This restores the signal's original input dynamic range. The noisy medium is effectively placed between the compressor and the expander so that the medium must pass a signal of substantially reduced dynamic range compared with its original state. The 2:1 compression and expansion ratios mean that the input dynamic range is cut in half for passage through the noisy medium. The dbx company claims that in terms of quietness and dynamic range, dbx noise reduction is markedly superior to any 16 bit PCM digital audio system.

The dbx 180A is a type I system that provides two channels of encode and two channels of decode processing. It is thus capable of processing both channels of a stereo signal or two independent single channel signals. Its primary use is to prevent tape hiss during recording. It will not remove hiss already present on a recording, nor will it reduce hiss on a tape that has been recorded without encoding.

Figure 3–7 shows the front panel of the dbx 180A. The rear panel of the rack-mounted dbx 180A contains two screw-mount barrier strips. One strip connects the encoder input portion to two channels

Figure 3–7   The dbx model 180A, front panel. (Courtesy dbx.)

of console outputs and the decoder output portion to two channels of console inputs. The second strip connects two recorder outputs to two decoder inputs and two recorder inputs to two encoder outputs.

The front panel features of the model 180A include a channel 1 and channel 2 in and out switch and LED indicator; depressing the button engages both the encoder and decoder of that channel only. The out position is a bypass with the inputs connected directly to the outputs. Next are the level adjust controls, which are recessed screwdriver-adjustable controls that adjust encoder and decoder gain for each channel. They are noncritical adjustments that are usually performed only once, when the unit is installed. Figure 3–8 shows the dbx 900 frame (A) and 911 module (B).

The dbx 900 series is a type I, nine-channel, noise-reduction, rack-mounted, modular system. Each 911 module is an independent encoder-decoder and can be connected to any of the tracks of a multitrack recorder, between the mike preamplifier of a console and the line level input of a tape recorder. Panel controls include a noise reduction (NR) in switch and recessed record and play level adjustment switches.

In addition to noise-reduction systems, dbx makes other signal-processing devices, such as compressor-limiters, de-essers (used to remove sibilance from an audio signal), noise gates (used to reduce background noise often found in remote feeds), and a real time spectrum analyzer.

Dolby Noise-Reduction System

The Dolby noise-reduction system (Figure 3–9) is the brainchild of Ray M. Dolby,

Figure 3–8 The dbx 900 frame *(left)* and 911 module *(right)*. (Courtesy dbx.)

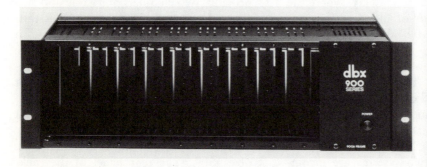

Figure 3–9 Two examples of the Dolby noise reduction system. (Courtesy Dolby Laboratories, Inc.)

Ph.D. The system includes an A type, which is used in professional applications, a B type, which is used in semiprofessional electronics applications, and a C type, which is commonly found in semiprofessional and consumer-grade open reel and cassette tape machines. There are also the Dolby SR (spectral recording) and HX (headroom extension) processes. Dolby SR is used primarily for master recording; Dolby HX is a dynamic noncomplementary (no decoder necessary) system used in the recording process to vary bias current relative to program frequency. Our description here is of the Dolby type A noise-reduction system, which can be used

with any audio recording or transmission system in which a signal is available both before and after a noisy segment. The Dolby A system uses a sophisticated form of compression and expansion, differing from conventional companding systems in two ways. First, it is a dual-path system, whereby louder signals that inherently mask noise pass through the system without processing, while companding (compression-expanding) takes place in a side-chain and affects only lower-level signals. This minimizes encode-decode errors and permits the use of effective signal overshoot suppression. Secondly, Dolby A is a multiband system where companding takes place at low level in four independent frequency bands. As a result, the presence of signals in one part of the audio spectrum cannot inhibit noise reduction in other parts, tending to eliminate the swishing and pumping sounds that sometimes are introduced as side effects by conventional companders. The Dolby A provides 10 dB of noise reduction from 30 Hz to 5 KHz, rising to 15 dB of noise reduction at 15 KHz. Tape noise, cross talk, low-level hum, and amplifier noise are all reduced, as is tape print-through. Additionally, harmonic distortion over the entire frequency range can be reduced because satisfactory signal-to-noise ratios can be achieved at lower recording levels.

Line Equalization

When a radio or television station obtains broadcast quality lines to carry music or other high-quality programming, the lines are equalized at both ends by the telephone company to compensate for line frequency loss due to line distance and whatever number of line amplifiers happen to be between the source of the line transmission and the station.

Equalization of the Listening Environment

To judge the aural quality of sound with discrimination requires two things: a reasonably accurate pair of ears and a satisfactory listening environment. This work environment is a combination of room acoustics and monitor loudspeakers. Neither the acoustics nor the loudspeakers should change the frequency response of the sound being judged by adding coloration. The listening environment should be as flat or uncolored as possible.

Since the room's acoustic design and the monitor loudspeakers may not always completely meet these criteria, we use equalization techniques to reshape the listening environment by adjusting the reverberation time of the room and speaker combination.

A pink noise generator's output is fed into the audio system, and its sound emanates through the loudspeaker monitors. The sound is then picked up by a microphone located in the listening environment and connected directly to the spectrum analyzer. If the curve that appears on the analyzer readout indicates nonlinear characteristics, then an equalizer, usually a parametric one, is inserted into the room's audio monitor system and adjusted to compensate for the room and speaker combination until the analyzer readout indicates as close to a flat curve as possible. With luck this is a one-time adjustment, unless the room parameters or the loudspeakers are changed. A cover plate over the equalizer controls ensures that they will not be changed after being set unless necessary.

There is increasing use of both passive and active notch filters in room equalization for an auditorium or hall. We know that a notch filter attenuates a small range of frequencies, typically ± 10 Hz from a center frequency. This removes often troublesome "ring modes" from a public address (PA) system in the auditorium without severe coloration of program audio. Ring modes are those frequencies at which the hall's natural period and reverberation time are sufficient to amplify a frequency without the PA system. If the PA system then further amplifies these frequencies, feedback will often result. A more traditional one-third-octave equalizer is usually placed after the notch filter to allow for further tailoring of the sound.

Because of ring modes, it is possible to have feedback even when a system does not seem to be operating overly loudly. By reducing the gain at the ring mode frequencies with a notch filter, it is possible to increase the overall PA gain as needed for entire program material before feedback occurs.

Excessive Equalization

A discussion of equalization would be incomplete without our noting some of the effects of overequalization. Already mentioned is the confusion the operator may experience in a complex music mix. Additionally, certain conditions could cause circuitry to become overloaded.

In the equalization of an audio system, the equalizers can be inserted—patched in or switched in—as either preequalization (ahead of a given stage of amplification) or postequalization (after the stage of amplification). The reader will remember that amplifiers are designed in stages, each stage with a specific number of decibels of gain. Each subsequent stage is designed to handle the previous stage's gain increase without distortion. The stage is then said to have sufficient headroom, and the equalization is intended to correct or compensate for a specific frequency deficiency. If preequalization is used, and it boosts the gain of specific frequencies in the signal flowing through that amplifier, then the gain boost will be added to the overall gain designed into the amplifier. When the frequency is low level, adding active equalizer gain causes no difficulty. If the boosted frequency is at a relatively high gain level to begin with, however, and it must be higher still, then the equalizer's additional gain boost, when added to the amplifier's normal gain, may overload that amplifier or the amplifier stages that follow. The additional gain may force those amplifiers to operate in a nonlinear fashion, or into saturation, on the peaks of the audio signal.

Our concern is with overall gain in a system. When active equalization is inserted into a system, the gain figures in that system have been changed, and some level of compensation is needed to prevent overload.

Limiter amplifiers are often employed to allow constant output control regardless of input level. Limiters operate by rapid attack against level peaks (1 msec) of perhaps 30 to 40 dB and a release time that can be set to several seconds.

The Eventide Harmonizer

The Eventide Harmonizer* is a multipurpose analog audio signal processor that provides pitch change, time compression-expansion, delay, reverberation effects, flanging, time reversal, and signal repeat.

It can change musical pitch one octave up or two octaves down. A front panel control sets the pitch ratio, which is displayed on a four-digit LED readout. The Eventide Harmonizer preserves harmonic relationships, and the pitch change output signal can be mixed with the input signal to create chorus and harmony effects. Like all pitch changers using time speedup-slowdown techniques, the Harmonizer produces "glitches" in the output during pitch change operation. Although the Harmonizer was designed to minimize the problem, the relative audibility and severity depends on the pitch ratio and also on the nature of the audio signal. This requires some operator experimentation before use on some types of music.

The Harmonizer's pitch change capabilities can be used for time compression-expansion by normalizing the pitch of recordings being played faster or slower than normal, altering the running time without editing. Three different frequency control outputs can directly vary the speed of professional tape recorders.

---

*The Eventide Harmonizer is manufactured by Eventide and is a trade name used by that company.

The Harmonizer will create a wide variety of delay and echo effects. Two outputs are available, each with variable delay up to 400 msec and variable feedback level. This allows multiple-length repeats. Since there are equalization controls for feedback, both bright and dead acoustic conditions can be created. Flanging, which is a short time delay with feedback but no pitch change, provides a hollow tunneling effect. In the time-reversal mode, a signal can be entered normally and outputted backward, providing unique effects. Signal repeat captures and repeats a 400 msec slice of program indefinitely.

SIGNAL PROCESSING IN DIGITAL AUDIO

Audio in the digital world makes processing easier. Equalization in a digital system is done automatically by computerlike circuitry.

A module of the digital console performs the processing functions that the audio operator defines for it, limited only by the console's computer processing power. The operator uses the console by choosing the order of the software routines that process the digital sound samples. A digital console is discussed in Chapter 4.

Digital processing equipment is currently found in such devices as pitch changers, time-compression devices, and digital delay and reverb units. One such combination device is the Yamaha SPX90II digital multieffect processor (Figure 3–10).

The SPX90II provides 11 different types of audio processing: delay, echo, modulation, gate, pitch, freeze, pan, vibrato, parametric EQ, reverb, and early reflections. It provides a full second of initial time delay for reverberation effects. Reverb is commonly associated with musical ambience, which is the totality of sound reflections in a hall, auditorium, or soundstage. These sound waves build and multiply into countless reflections. The SPX90II provides early reflection, an effect that recreates the sound wave immediately after an original sound. Early reflection is akin to the "slapback" effect used in vocal and percussion instrument recording, adding "presence" to the signal. The SPX90II creates delay, commonly used in recording.

It produces independently variable left and right channel stereo signal delays for a "doubled" sound. The SPX90II creates echo, which is similar to delay and gives added dimension to both instrumental and vocal music. Although reverb creates partial sound reflections and delay produces

Figure 3–10 The Yamaha SPX90II digital multi-effects processor. (Courtesy Yamaha.)

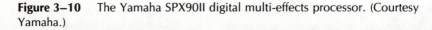

a limited number of signal repetitions, echo can produce limitless signal repetitions.

With SPX90II, modulation effects are produced by periodically varying the amplitude, the frequency, or the delay time of an input signal. Stereo flanging, chorus, stereo phasing, tremolo, and symphonic effects are all available.

The SPX90II has an auto pan program that automatically pans the signal between left and right stereo channels. Pan direction, speed, and depth can be programmed. The SPX90II produces vibrato effect, where minute pitch variations are adjustable over a wide range. It also has a noise gate circuit that allows a short segment of a longer signal to pass or be shut off. It can also pass only signals above a specific level, or it can reverse the gating effect so that the gain increases gradually after the gate is triggered. The gate may also be triggered by a foot switch. The SPX90II can combine gating with reverb, and it can change pitch in four different programs, making it possible to produce harmonizer and chorus effects.

It has freeze programs, which permit recording of up to 2 full seconds in program random access memory for playback as required. SPX90II has compression, a process where an input signal's dynamic range is reduced, a low-input signal level is increased, and a high-input signal level is reduced. The SPX90II has parametric equalization. It can preset 30 effects in advance of use and has 60 user-programmable memory locations.

AUDIO SIGNAL DISTORTION

We have discussed the methodology and equipment used to correct audio distortion without really considering distortion itself. Distortion is usually caused by one of two reasons. The first is operator mis-adjustment of the controls, which allows overly high audio levels and thus overloaded circuits to clip the signal tops. The second reason for distortion is any minor change in equipment circuitry. Changes are caused slowly by age, heat, or minor variation from design parameters; they modify the otherwise pure signal as it passes through the circuitry.

Distortion can be identified by type, with the most prevalent being intermodulation distortion (IMD), total harmonic distortion (THD), phase distortion, and frequency distortion. All analog audio systems have some degree of all types of distortion at all times. A well-designed audio system keeps distortion to a minimum (less than 0.01%). Good operational practices and sterling equipment maintenance will help keep distortion to an absolute minimum.

Intermodulation Distortion

The most often discussed type of distortion in the audio literature of recent times is IMD. It occurs when part of a system begins to operate in a nonlinear fashion, that is, when the tops of the audio signal peaks are being clipped. This is sometimes called *flat topping*. It causes one lower frequency component of the audio signal to modulate a higher frequency component of that same signal. The modulation produces still an additional and unwanted component of the signal, which changes the original signal.

Phase Distortion

Every frequency component of the audio signal should pass through an amplifier at precisely the same time for distortion-free amplification. If part of the audio signal (some of the frequencies) are caused by anomalies in the amplifier to lead or lag the rest of the signal by perhaps microseconds, then phase distortion will occur. A capacitor that is slightly changed in value by heat or age could literally delay the passage of a high frequency component of the signal—one note—just enough to change the relationship of that note to the rest of the music. Phase distortion is also known as *time distortion.*

Frequency Distortion

When the original design parameters that affect an amplifier's frequency response are altered by the failure, or modification, of an electronic part in that amplifier, each specific frequency in the audio spectrum passing through that amplifier will not be amplified equally, causing frequency distortion. Frequency distortion often arises when two or more electronic devices whose output-to-input impedances are badly mismatched are patched or matrixed together.

Total Harmonic Distortion

THD is created in an audio system when the circuitry itself inadvertently generates, through minor oscillation, multiples or submultiples of a particular frequency passing through the system. These multiples, called harmonics, are added to or subtracted from the original frequencies, causing signal change.

Other miscellaneous causes of distortion in an audio system can include cold solder joints in the wiring, loose or sloppily wired connections, corrosion (oxidation) on a plug, jack, or other mechanical contact surfaces, mechanical-acoustic infusion of turntable or tape machine motor rumble picked up by the device's transducer, and radio frequency interference from transmitters close to the audio system.

REVIEW QUESTIONS

1. State the upper and lower limits of the audio spectrum.
2. What is an audio filter? An equalizer? What can an equalizer not do?
3. Describe a graphic equalizer and a parametric equalizer.
4. What is a beeper call? What is pink noise?
5. What does a low-frequency extender system do?
6. Describe a compander. What does it do?
7. How do we equalize a listening environment?
8. Describe and explain four types of audio distortion.
9. What is a limiter amplifier?

4

PROFESSIONAL CONTROL BOARDS

We will now examine in some detail a variety of professional control consoles. Note that this text does not promote any manufacturer's products or, for that matter, express the author's preferences. It dissects these products impartially to acquaint the newcomer to the industry with a sampling of what equipment he might be asked to use on the job. The importance of this exercise is to show that all consoles or boards have many similarities but some vast differences.

CONSOLE CONSTRUCTION

Consoles are either hard wired in one complete unit, where components must be unsoldered and mechanically disconnected for replacement, or made modular, that is, basic parts are on circuit boards called modules, which unplug from each other for disassembly.

Modular consoles have a covered metal frame, which holds the printed circuit board modules. The modules plug in to a "motherboard," which carries all the interconnections between modules, busses, routing paths, and power supply facilities. The input modules containing input-level setting switch, pot, key, preamplifier, assignment switching, and signal processing plug into this large circuit board, as do

the modules with a monitor amplifier, a program amplifier, a line amplifier, and a distribution amplifier.

The modular console's mounting frame can often be ordered in several sizes. Some have blank panel covers over empty input or other facility spaces to incorporate projected expansion.

Console maintenance, which formerly required extensive down-time, can be accomplished rapidly in modular consoles by unplugging a defective module and replacing it with a functioning one.

Consoles are connected to the audio world with "shielded pairs" of audio cable. Most consoles that are mounted permanently in control rooms have "barrier strip" connectors, multiple-screw-connector strips to which the outside connections are made with spade lugs at the ends of the wire pairs, under the screws. Many consoles have provisions for effects loops. The TRS connector, which is described in Chapter 5, is used in an unbalanced state, to insert the effects device within an input or at the group output. Typically, the ring of the TRS connector is used to send audio to the effects device. The tip is used to return audio from the effects device. The insert point usually employs a -10 dBv level, is bridging, and appears postfader (after the fader but before the VU meter or the monitor circuits). The insert point can also be used as a line level input within the input channel by wiring a TRS plug for tip and sleeve. If an insert point is labeled *Insert* or *Effect Loop,* only TRS-type plugs should be used to avoid short-circuiting the ring connection at the console, with possible damage.

CONSOLE SPECIFICATIONS SHEET

The console manufacturer virtually always includes a specifications, or spec, sheet in the advertising material printed about a console. The spec sheet includes the number of inputs and outputs; input and output impedances (Zs); the equivalent input noise (EIN) figure, which is a performance rating; available input and output levels; physical dimensions and power requirements; amplifier responses, distortion figures; and signal-to-noise ratios.

All consoles made today are solid-state devices. Before the advent of the chip, which performs all console amplification, the vacuum tube was used.*

TYPES OF CONSOLES

Broadcast Electronics, Inc., Consoles

We will examine and describe three consoles made by Broadcast Electronics: models 4M50A, 10S250, and 10S350.

---

*A historical note is of interest here: At small, inexpensively run stations in the American hinterlands, called quarter kilowatters because the FCC licensed them for 250 W of AM power, something strange often happened to new consoles. When the FCC proof of performance tests were performed, the vacuum tubes that came with the console from the factory, and were the best tubes that the manufacturer could buy, were left in the console. When the tests were over, the tubes were removed and carefully boxed to be saved for the next test. They were usually replaced with cheaper tubes from the local radio store!

Model 4M50A

The 4M50A is a basic nonmodular monaural console. It has a stereo counterpart, the 4S50A. Note that the console, illustrated in Figure 4–1, is similar to the hypothetical four-input-channel console described in Chapter 2 and pictured in Figure 2–7. The open view, which is of the stereo model, shows the front panel tilted down, the two meters on the right, the barrier strip at the upper left, the power supply at the upper right, and the cue speaker below the power supply.

The 4M50A has four input channels, each one selectible to two input sources, for a total of eight available inputs. All four input channels have preamplifiers, and all the inputs may be strapped (internally jumpered) for high- or low-level input impedance. Input 1 has normal muting capability for mike use. Each input has a rotary pot with detent cue switch at the closed pot position. There is no audition buss or channel-off key switching on this console. Located above the input selector keys, which select input A or B, are the master pot, the monitor pot, and the cue pot, followed by the phones pot and its selector switch and by the VU meter. The monitor pot controls a monitor amplifier; the cue pot controls a cue amp and the built-in cue speaker, which is also muted when mixer number 1 is on. Muting is available for the other inputs and may be strapped as needed. The headphone jack, located at the lower right of the console, is switchable to the program buss, the cue buss, or an external source. This basic console might find use in a newsroom or other remote location for cut-ins to the main studio. Figure 4–2 shows a functional diagram of the 4M50A.

From the spec sheet:

1. Program channel: input impedances-levels (strappable); low mode: 150 ohm balanced, −65 dBm nominal,

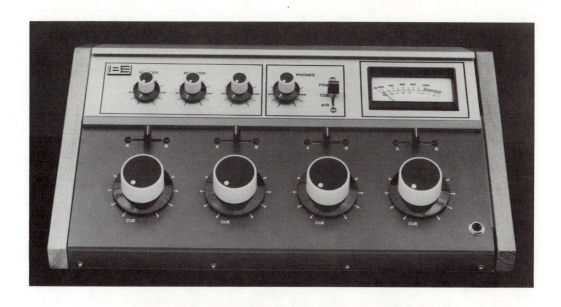

Figure 4–1   The Broadcast Electronics 4M50 console, front *(top)* and open *(bottom)* views. (Courtesy Broadcast Electronics, Inc.)

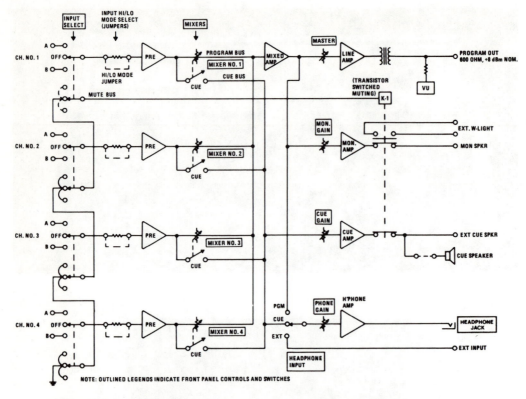

Figure 4–2 The Broadcast Electronics 4M50 console, functional diagram. (Courtesy Broadcast Electronics, Inc.)

−45 dBm maximum; high mode: 20 Kohm balanced bridging, −20 dBm nominal, 0 dBm maximum.

2. Frequency response-distortion: ±2 dB −0.5%, from 30 Hz to 20 KHz.

3. Signal-to-noise ratio: 65 dB (unweighted) below +8 dBm output, −50 signal to any low-level input.

4. Overall gain: 90 dB minimum.

5. Output impedance-level: 600 ohm balanced, +8 dBm for 0 VU meter reading, +16 dBm maximum.

Model 10S250

Model 10S250 (Figure 4–3) is a ten-input-channel stereo console with four line outputs. In its mono configuration it is the 10M250, with two line outputs. Its

ten inputs accommodate 20 input sources by a matrix of input select push switches located above the rotary input pots and keys. The keys each select audition, off, or program. To the left of the input selectors is the cue speaker, and to the right are the talkback push switch and the selector switches for two muted studio areas and one nonmuted area such as a station's lobby or auditioning room. Following those switches are the two VU meters. One meter can be switched to each of the program and audition busses. The headphone jack, located at the lower right, is fed from an earphone amplifier with level control that is switchable to program, audition, or cue. The monitor amplifier has a level control and is also switchable to program, audition, or cue. The cue amplifier is fed from detent positions on each

Figure 4–3   The Broadcast Electronics 10S250 console. (Courtesy Broadcast Electronics, Inc.)

input pot at the pot's closed position. Figure 4–4 shows a functional diagram of the 10S250.

From the spec sheet:

1. Both program and audition channels have a choice of input impedances and levels: low mode: 150 ohm balanced, −65 dBm minimum, −38 dBm maximum. High mode: 54 Kohm balanced, bridging, −20 dBm minimum, +20 dBm maximum.
2. Frequency response: ±1 dB, from 30 Hz to 20 KHz.
3. Distortion: 0.05% or less IMD and THD at +18 dBm output, from 30 Hz to 20 KHz.
4. Signal-to-noise ratio: Noise (unweighted) 70 dB below +18 dBm output with −50 dBm signal into any low-level input, 20 KHz bandwidth.
5. Overall gain: 105 dB.
6. Output impedance-level: 600 ohm balanced, +8 dBm for zero VU meter reading, +18 dBm output capability.

Model 10S350 and 10M350

Model 10S350 stereo, and 10M350, mono, (Figure 4–5) are ten-input-channel consoles with 22 selectable input sources. They have two-output-channel capability in mono and four outputs in stereo. Unlike the other two BEI consoles, these consoles use vertical faders. Each of the input faders 1 through 8 accepts one of two input sources. Input faders 9 and 10 each accept one of three input sources. Input selection is by interlocked push buttons above the vertical pot and key. The keys select, vertically, program, off, or cue. Cue gain is independent of fader setting. The individual inputs may be preset to either mike- or line-level input impedance. The outputs of these input mixers are individually switchable to either of two mixer busses and hence to the output channels. Console monitoring of either of the two program channels or of the cue channel is available to speakers and earphones. Talkback to the studio is available using the cue speaker as a microphone transducer. Speaker muting is normally assigned to only input-channel mixers 1 and 2 but may be expanded to include any of the other inputs. Figure 4–6, the functional diagram of the mono version of this console, shows input channels 1 through 3, since channels 3 through 8 are identical, and then either channel 9 or 10, since these two are identical.

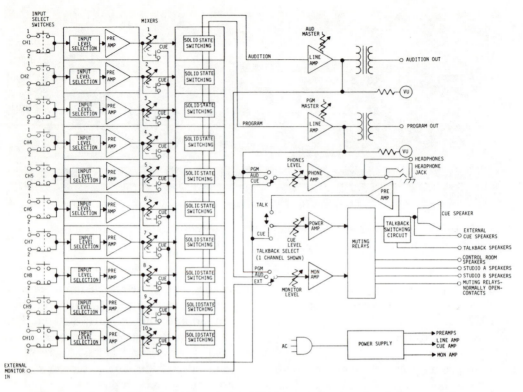

Figure 4–4   The Broadcast Electronics 10S250 console, functional diagram. (Courtesy Broadcast Electronics, Inc.)

From the spec sheet:

1. Input impedances-levels (switchable): low mode: 150 ohm balanced, −65 dBv nominal, −38 dBv maximum; high mode: 54 Kohm balanced bridging, −20 dBv nominal, +20 dBv maximum.

2. Frequency response: +0, −1 dB, from 30 Hz to 20 KHz.

Figure 4–5   The Broadcast Electronics 10S350A console. (Courtesy Broadcast Electronics, Inc.)

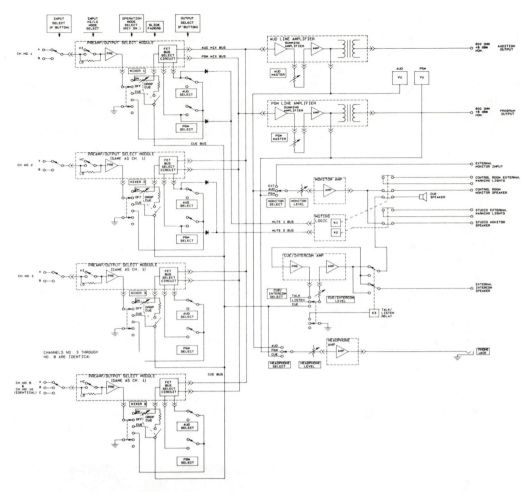

Figure 4–6 The Broadcast Electronics 10M350A console, functional diagram. (Courtesy Broadcast Electronics, Inc.)

3. Distortion: 0.075% IMD and THD, from 30 Hz to 20 KHz at +18 dBm output.
4. Signal-to-noise ratio (unweighted): 68 dB below +18 dBv output, −50 dB input over the 20 KHz bandwidth.
5. Output impedance-level: 600 ohm balanced, +8 dB for zero VU meter deflection, +18 dB capability.

ADM Technology, Inc., Consoles

ADM consoles are completely modular and permit a particularly close look at

their modules. ADM makes the ST series II consoles. Primarily designed for radio, the series consists of four basic units, each identical except for its number of input modules. The ST 100 II is a ten-input console; the ST 160 II, a 16-input console; the ST 200, a 20-input console, and the ST 240, a 24-input console. All four consoles have three stereo outputs and one monaural output.

We will first look at some of the ST series modules. In this examination we will see that the purchaser of a modular console has a number of choices to make, even after selecting a specific console make and model. The choices include number of inputs and how many inputs shall have processing facilities; whether the input modules are mono or stereo, mike or line; the number and types of busses to which the inputs shall be routed; cuing and audition facility requirements; and the impact of future requirements on the console.

Module 4711

The 4711 Slidex voltage-controlled amplifier (VCA) attenuator module (Figure 4–7) has a vertically operated pot that provides the superior tracking of a VCA chip and noise-free attenuation. It is separately constructed from, but electronically part of, the input module on the series II consoles. It has a cue detent switch incorporated at the bottom of the slide and a push-type illuminated module on-

off switch (key). This key controls all the logic circuits for microphone switching and studio–control room monitor muting. The key is two-color coded and indicates by illuminated color when the module is on or off. A separate cue switch with associated LED indicator is enabled only when the module is off, and it performs a parallel function to the Slidex cue detent position.

Input Module

The input module (Figure 4–8) comes in monaural and stereo versions, in three-mike- and three-line-input versions. These modules can accept one of two selectable input sources, called A and B. Each selector has an associated LED indicator for the operator's visual information. A mike-level input can select either mike A or B,

Figure 4–7 The Slidex attenuator module, front *(left)* and side *(bottom)* views. (Courtesy ADM Technology, Inc.)

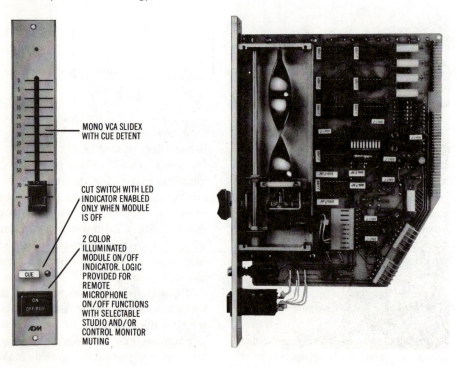

MONO VCA SLIDEX WITH CUE DETENT

CUT SWITCH WITH LED INDICATOR ENABLED ONLY WHEN MODULE IS OFF

2 COLOR ILLUMINATED MODULE ON/OFF INDICATOR. LOGIC PROVIDED FOR REMOTE MICROPHONE ON/OFF FUNCTIONS WITH SELECTABLE STUDIO AND/OR CONTROL MONITOR MUTING

**2716 MONO MICROPHONE
INPUT MODULE**

2 POSITION
MICROPHONE
INPUT SELECTOR
SWITCHES WITH LED
STATUS INDICATOR

CONTINUOUSLY
VARIABLE INPUT
GAIN CONTROL WITH
30dB OF RANGE FOR
INPUTS OF −60 TO
−30dBv (REF. .775)
WITH REMOVABLE
KNOB AND SHAFT

AUXILIARY MODULE
OUTPUT WITH
PRE/POST FADER
SELECTION AND
LEVEL CONTROL

STEREO PAN POT
WITH MONO/PAN
SWITCH

PROGRAM/AUDITION
OUTPUT ASSIGNMENT
SWITCHES
(ILLUMINATED)

**2717 MONO MICROPHONE
INPUT MODULE**

2 POSITION
MICROPHONE INPUT
SELECTOR SWITCHES
WITH LED STATUS
INDICATOR

INPUT GAIN CONTROL
WITH KNOB AND
SHAFT REMOVED

STEREO PAN POT
WITH MONO/PAN
SWITCH

PROGRAM/AUDITION
OUTPUT ASSIGNMENT
SWITCHES
(ILLUMINATED)

(SIDE VIEW)

Figure 4–8 Input module, front *(top)* and side *(bottom)* views. (Courtesy ADM Technology, Inc.)

and, in the line configuration, line A or B. Below these switches on the module is an input-attenuator selector that can be set from −60 to −30 dBu depending on the input impedance of the mike or line chosen. Below this is the auxiliary module output control. Each type of input module has a facility for a totally independent

auxiliary output, which is selectable from either the preamp output (PRE) or module (POST) output. The AUX output has its own level control located concentrically with the switch function. In the PRE switch position it is totally independent, while in the POST position the AUX output follows the module's Slidex attenuator and signal processor, if any. Below the AUX control is the stereo panpot and, concentrically, the mono-pan switch. The module is assigned (routed) to a program or audition buss by its output-assignment switcher, which is illuminated when in operation. The right edge of the side view (see Figure 4–8) is the plug-in connector edge.

A line-input module is selectable to one of two high-level sources labeled A or B at the top of the module, and the attenuator selector below can be set to an input level of from −12 to +8 dBu depending on the gain of the incoming line. The line-input module has a stereo balance control to trim inequities in the stereo program source and a concentric switch to select left, right, stereo, or mono mode for the module. Similar to the mike-level module, there are program and audition buss assignment switches. Associated with both the mike- and line-input modules are two eight-position preselect matrices located on separate modules but standard on ST series II consoles (Figure 4–9). Each of these matrices may be assigned to a line- or mike-input module by a plug-in connector. A switch matrix, similar to the switch bank on a touchtone telephone, permits assignment of a combination of switches.

The output buss assignment of the input modules is available to both audition and program master busses through illuminated switches and may be fed individually or simultaneously as required. There may be up to 11 ADM signal-processing modules mounted in the console, consisting of equalizer, sound effects filter, noise suppressor, and limiter-de-esser modules in the input circuitry.

Equalizer Module

The equalizer module (Figure 4–10) is a four-band device. It has a high-frequency section of 10, 12.5, and 15 KHz; a mid-

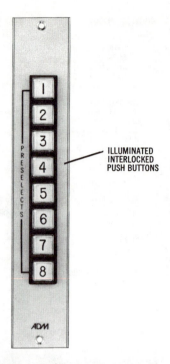

Figure 4–9   Eight-position stereo line preselect module. (Courtesy ADM Technology, Inc.)

Figure 4–10   Equalizer module. (Courtesy ADM Technology, Inc.)

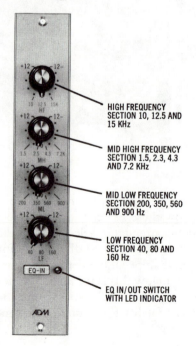

high-frequency section of 1.5, 2.3, 4.3, and 7.2 KHz; a mid-low-frequency section of 200, 350, 560, and 900 Hz; and a low-frequency section of 40, 80, and 160 Hz.

There is an EQ in-out, push-off–push-on switch with LED indicator. The curves in Figure 4–11 describe its individual equalization functions.

Figure 4–11   Equalizer module curves. (Courtesy ADM Technology, Inc.)

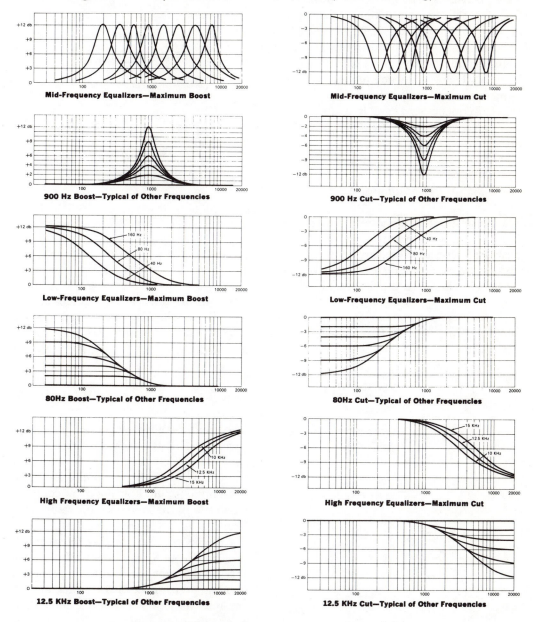

H.F. EQUALIZER FREQUENCIES: 10, 12.5, 15 kHz ±2, 4, 6, 9, 12 dB
H.M.F. EQUALIZER FREQUENCIES: 1.5, 2.5, 4.3, 7.2 kHz ±2, 4, 6, 9, 12 dB
L.M.F. EQUALIZER FREQUENCIES: 200, 350, 560, 900 Hz ±2, 4, 6, 9, 12 dB
L.F. EQUALIZER FREQUENCIES: 40, 80, 160Hz ±2, 4, 6, 9, 12dB

High- and Low-Pass Filter Module

The filter module (Figure 4–12) was specifically designed for sound effects. The high-pass section comprises selection of 50, 70, 100, 150, 200, 300, 500, 700, 1000, and 2000 Hz and an off position. The low-pass section comprises selection of 700 Hz and 1, 1.5, 2, 3, 4.5, 6, 7.5, 10, and 12.5 KHz plus an off position. Both sections may be used simultaneously. There is an attenuation rate of approximately 18 dB per octave at the selected frequencies. There is a push-on–push-off switch with attendant LED indicator.

Noise-Suppressor Module

The noise-suppressor module (Figure 4–13) is a true gain expander with continuously varying gain. There are no threshold clicks or pops, nor is the ear aware of the threshold. There are threshold and decay

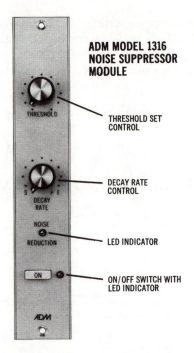

Figure 4–13   Noise-suppressor module. (Courtesy ADM Technology, Inc.)

rate controls and a variable-intensity LED to indicate when the unit is on. A push-type on-off switch and LED indicator are included.

Limiter-Compresser-De-esser Module

The limiter-compresser-de-esser module (Figure 4–14) is both a limiter compressor and a de-esser. Each function, although independent, is contained within the same module housing. The compression ratio of the limiter-compressor varies from 1:1 to 20:1. The ratio changes from 1:1 to the selected value as signal power increases from 8 dB below to 8 dB above threshold. Front panel controls permit individual adjustment of the attack and decay rates as well as threshold and gain.

The de-essing function, when activated, limits sibilance to natural levels with no adjustment required.

Figure 4–12   High- low-pass filter module. (Courtesy ADM Technology, Inc.)

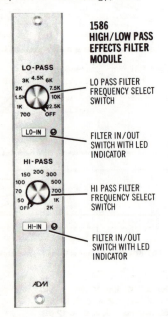

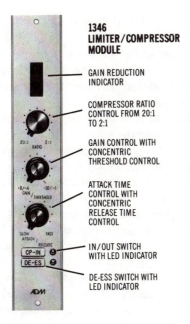

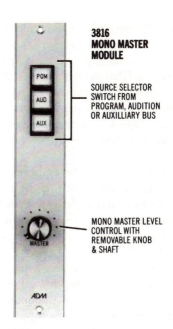

Figure 4–14   Limiter-compressor module. (Courtesy ADM Technology, Inc.)

Figure 4–15   Mono master module. (Courtesy ADM Technology, Inc.)

The stereo master output modules are not pictured in this text because they are externally blank module panels with only a single master-level control. The series II consoles each employ three stereo master modules and one monaural module. The monaural master module (Figure 4–15) is selectable through its illuminated switch matrix to the audition, program, or auxiliary busses, where it combines the stereo signal to monaural before the user selects the master stereo module, permitting totally independent operation.

The cue functions on the Slidex modules feed a cue buss to a cue amplifier with level control located in the cue-talkback module. This module is standard on all ST series II consoles. The cue amplifier feeds a cue speaker located in the meter turret housing at the top of the console. In addition, cue output appears at an output connector for use with an external cue speaker and at the earphone jack. It is accessed through the monitor-cue switch and

muted to the cue speaker whenever a control room mike is live. The headset cue signal is never muted. The talkback portion of the cue-talkback module is provided with a level control for talkback output. Associated with this module is a talkback mike and a push-to-talk switch located in the console meter turret. The talkback signal is normally fed over the studio monitor buss, overriding any signal being fed to the studio monitor speakers.

The control room headphone jack is located on the right front section of the console. It is selectable through the monitor cue switch to hear either program, as selected on the control room monitor selector, or output of the cue buss. Headphones permit the operator to hear program while a control room mike is live.

The series II console has four VU meters. Two, designated program, are dedicated to the console program outputs. The other two meters are designated utility and are switch selectable to the audition buss, monaural buss, and auxiliary buss and to two spare meter inputs designated external 1 and 2. This illuminated selector matrix is located on the lower left of the console, beneath the input signal processor area. Zero VU on the meters is equivalent to a nominal +8 dBu line output level.

Each console may be equipped with a 60-minute timer, which is designed to count up or down as desired. Preset times may be entered into the display through the timer's second and minute advance functions.

The console is provided with a power supply capable of supplying all necessary power with more than adequate safety factor. A built-in alternate power supply will provide complete power in the event of failure. LED status indicators for the power supplies are located in the meter turret housing.

Model ST 100 II Stereo Console

Pictured in Figure 4–16 is the ST 100 II stereo console. Having looked internally at its modules, we now turn our attention to the console itself.

We will first examine the layout of its face, noting the various modules' mounting positions, and then we will look more closely at a panel layout drawing.

We see first that the console has a sloping front and that it is designed to sit on, or be attached to, a desktop. Its upper portion or turret contains, from left to right, the manufacturer's logo with the power supply status LED indicators below. Next is the cue speaker, followed by the two utility VU meters, the talkback mike, the two program VU meters, and the optional timer.

The sloping front panel has, from left to right, six blank expansion panels, into which equalizer modules could be in-

Figure 4–16   The ADM ST 100 II stereo console. (Courtesy ADM Technology, Inc.)

stalled, below which is the utility meter selector. Following are the two eight-position line preselect modules, top and bottom. Then come the ten Slidex attenuator modules, and above each are the mike- or line-input modules. To the right of the input modules are additional blank panels and the four master output modules. Below these is a horizontal module plate containing the timer controls. To the right of the master modules, the upper right module is the talkback-cue, with the push-to-talk switch, and below it are the control room and studio monitor selector matrices and level controls. The earphone jack is on the extreme bottom right.

Model ST 160 II Stereo Console

Figure 4–17 shows the same console with 16 inputs and, on the extreme left, equalizer modules. Figure 4–18 illustrates a panel layout drawing of the ST series II consoles, and Figure 4–19 presents a block diagram of these consoles.

From the ADM spec sheet:

1. Frequency response: with no equalization, measured at any output level up to clipping, ±1 dB, 20 Hz to 20 KHz, reference 1 KHz.
2. Distortion: the THD at +24 dBm or lower at 1 KHz will be less than 0.07% and will not exceed 0.15% THD over the band 100 Hz to 20 KHz at +24 dBm or lower.
3. Maximum output level: the clipping level at any output when terminated in 600 ohm shall be +27 dBm from 30 Hz to 20 KHz.
4. Noise: the EIN of any microphone shall be lower than −125.5 dBu referred to a 250 ohm impedance measured on an average response meter. Any line-level input (+8 dBm reference) to any output channel (+8 dBm reference) will exhibit a maximum noise of −72 dBm (SNR 80 dB). All

Figure 4–17   The ADM ST 160 II stereo console. (Courtesy ADM Technology, Inc.)

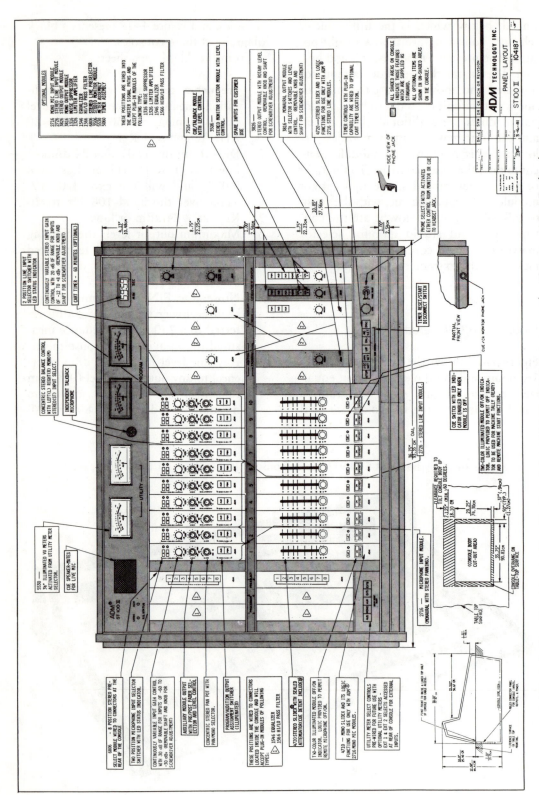

Figure 4-18 Panel layout line drawing of the ADM ST 100 II stereo console. (Courtesy ADM Technology, Inc.)

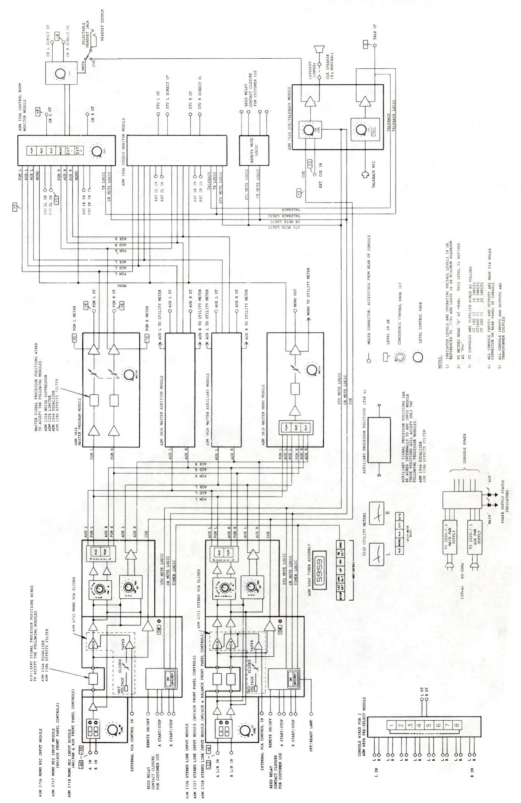

Figure 4–19 Block diagram of the ST series II consoles. (Courtesy ADM Technology, Inc.)

noise measurements are based on a bandwidth of 20 Hz to 20 KHz.

5. Crosstalk: better than 72 dB measured between adjacent channels at normal operating levels over the 100 Hz to 10 KHz band.

Studer Revox Consoles

Studer Revox is a Swiss company that makes tape recorders and system controllers as well as consoles. We will look at three of the Studer consoles—the series 900, the series 961-962, and the series 963.

900 Series

The 900 series (Figure 4–20) is available in a model 901, with up to 13 inputs and 4 output busses–channels; the model 902, with up to 28 inputs and 8 output busses–channels; and special-order consoles (model 904) that can be supplied with up to 50 input channels and 24 master busses for multitrack mixing.

Input modules for the 900 series consoles are available in two basic versions, A and B. Version A, the standard type, features a microphone presence-absence filter and parametric equalization sections. The input module can be driven by either mike- or line-level signal and is op-

Figure 4–20 The Studer Revox series 900 console. (Courtesy Studer Revox America, Inc.)

tionally available with a mike input alone. The B version is a simplified module, with fewer filter, EQ, and input selection facilities. Type B modules are also available with only stereo high-level inputs with mono or dual faders. Figure 4–21 shows the Studer 900 input and output modules.

The A version mono-input module has four switchable inputs, rotary concentric gain match controls, and high- and low-level selectors (generator, mike, tape, and line). It also has an audio oscillator (generator) input, an overload peaker, phantom power switch, phase reverser, filter and EQ sections, EQ in-out switch, four AUX controls, pan in-out switch and control, mute switch, and master selector, which chooses master buss 1 through 4 (or 1 through 8 on the larger console version).

The stereo input module with EQ provides a choice of one of four pairs of inputs, overload peaker, mike and line gain match controls with concentric switching, phantom power on-off, left and right channel inversion, stereo spread control, filter and EQ section, a four-output AUX section, stereo balance control and switch, mute switch, and master buss selection. This module has dual faders and can be ordered with VCA faders. VCA faders indirectly influence the control of gain by changing the preamp's operational mode between class A and class AB amplification rather than directly, as standard resistive faders or pots do.

The output modules have dual bar graph displays, with VU and PPM characteristics. Each display has 200 luminescent segments. There is a dual master output selector that assigns each submix group to one or more master outputs; four pairs of AUX controls with a panpot for each set of pairs; tape, solo, and mute switching; and dual master faders.

Additionally, the 900 series consoles have AUX output master modules; monitor mixer modules; monitor modules that provide studio monitoring, talkback, PFL, and solo; and control room monitoring.

From the spec sheet:

1. Input Z: mike: 1.2 Kohm, line: 10 Kohm.
2. Output Z: 50 ohm.
3. Frequency response: 31.5 Hz to 16 kHz, $+0.5$, -1 dB.
4. Distortion: 31.5 Hz to 16 kHz, $>0.1\%$.

960 Series

The Studer 961 and 962 consoles (Figures 4–22 and 4–23), part of the 960 series, are lightweight portables with input and output connectors on XLR-3s, bantam jacks for the insert points, and 50 pin connectors for additional, separate monitor connector panels on the rear of the console. The 961, fully equipped, with case and cover, weighs 55 lb; the 962 weighs 75 lb. The 961, with up to 14 modules, and the 962, with up to 20 modules, share the following features: internal AC power supply with power cable on rear; optional operation by external battery supply; a plug-in meter panel attachable to the console's plug-in handle; optionally VU or PPM meter configuration; a master channel limiter compressor module; switch selectable phantom power on all mike-input channels; and peak level indicators (LED) on all input and high-level channels. There are two small meters for viewing limiter-compressor gain reduction, two small meters for the AUX channels, and an optional full-sized correlation meter that correlates between the two main gain meters. The 961s have two output or master channels, and the 962s have four.

The mono input modules for the 961

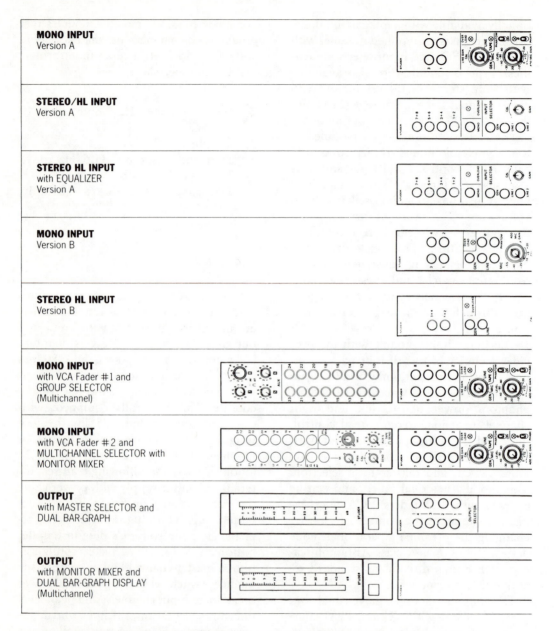

Figure 4–21 The Studer Revox 900 input and output modules. (Courtesy Studer Revox America, Inc.)

and 962 consoles (Figure 4-24) are identical except for assignment channel switching and the associated panpot. The top of the module strip has an overload peaker, followed by a concentric input four-step sensitivity switch and a mike-, line-, generator-off switch; phantom power, phase reversal, filter, and simple shelving equalizer with bypass; two AUX output controls and two or four

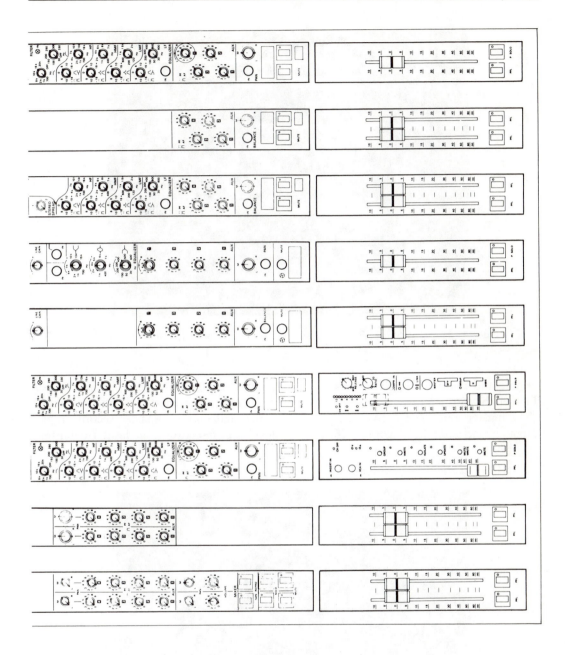

output-channel selectors with panpot; and a mute switch with LED, PFL switch, and linear vertical fader.

The stereo high-level input modules (Figure 4–25) are identical, except that one is available without an equalization section. The top of the module has an overload peaker; input gain balance pot; equalizer section in one model; two AUX output controls; channel select switch and

Figure 4–22   The Studer Revox series 961 console. (Courtesy Studer Revox America, Inc.)

panpot; and mute with LED indication, PFL switch, and fader.

The master modules for two and four output-channel consoles (Figure 4–26) are similar except for the channel select switching. Each master module has three functional entities: master section, high-level input section, and limiter-compres-

Figure 4–23   The Studer series 962 console. (Courtesy Studer Revox America, Inc.)

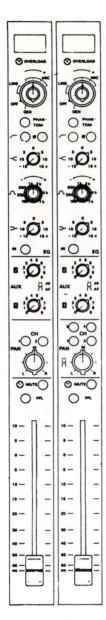

Figure 4–24   The 960 mono input modules for two- and four-output consoles. (Courtesy Studer Revox America, Inc.)

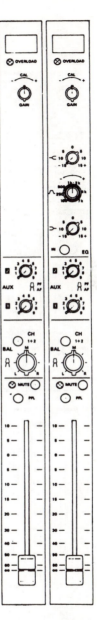

Figure 4–25   The 960 stereo high-level input modules. (Courtesy Studer Revox America, Inc.)

sor section. The limiter-compressor is inserted by pushbutton either as a line protector limiter in the master channel or as a compressor into any input channel. The link button couples the limiter and

compressor for stereo control. Each has LED indicators at the top of the master module, followed by operational controls. Next is the additional high-level input pot and PFL facility, two AUX

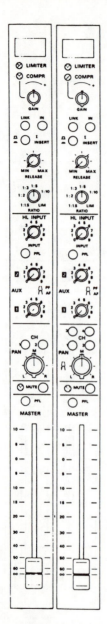

Figure 4–26 The 960 master modules for two- and four-output consoles. (Courtesy Studer Revox America, Inc.)

Figure 4–27 The 960 control room monitor module. (Courtesy Studer Revox America, Inc.)

controls, channel select and panpot, mute button and indicator, PFL button, and master fader.

In the control room monitor module (Figure 4–27), nine different monitoring sources may be chosen with interlocking

push buttons and their level controlled with a pot. Volume imbalances caused by room architecture or monitor speakers can be compensated for with the balance pot. Stereo sources can be monitored in mono mode using the mono button. Output meters 1 and 2 can be selectively connected to master outputs 1 and 2 or paralleled to the monitor loudspeakers with the meter button. With the PFL to monitor button the built-in speaker may be disabled. Alternately, as soon as a PFL button on an input is pressed, the signal to be monitored is interrupted and the selected PFL signal is switched to the monitor speakers. This does not interfere with a program in progress. When all PFL buttons are released, the selected monitor program becomes audible again. Three pushbuttons follow for studio signaling, with their status indicated on three LEDs on the meter panel. A red signal, ON AIR, over the studio door, is illuminated if one or more mike channels are on. A green attention signal is activated in the studio as long as this button is depressed. A yellow call signal is illuminated as a warning signal when this button is pressed.

The AUX, communication, and talkback module (Figure 4–28) has a connector and switch for a miniature console utility light, switching for a five-frequency test generator, a built-in mike and switch for talkback, two AUX send (output) controls, and talkback switching for slate and studio with gain control. Slate, a film term, describes a tone burst that indicates the film is up to speed.

Studer provides a unique level diagram (Figure 4–29) that shows an input-module-to-master-output-module functional diagram along the top, with corresponding levels in the console graphed in dBu below.

From the spec sheet:

1. Inputs: mike Z: 1.6 Kohm, line Z: 10 Kohm.
2. Frequency response: 31.5 Hz to 16 kHz, +0.5, −1 dB.
3. Signal-to-noise ratio: >92 dB.
4. THD: <0.03%.

Model 963 (Figure 4–30) is a multi-input console. Its minimum configuration

Figure 4–28   The 960 AUX, communication, and TB module. (Courtesy Studer Revox America, Inc.)

is 12 input modules, and its maximum is 60 input modules, 4 or 8 submix groups, 4 AUX masters, and 4 master outputs. It was designed for use in remote broadcast vans and for sound reinforcement in theaters, multi-input mixdown, and audio postproduction. Since the modules are ex-

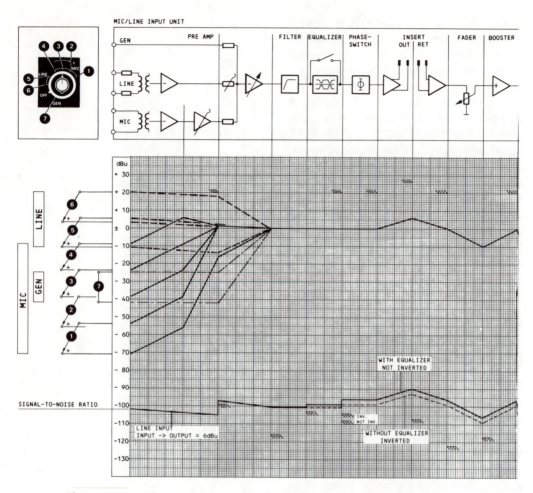

Figure 4–29   The 960 level diagram. (Courtesy Studer Revox America, Inc.)

panded versions of the 961 and 962 modules, we will only describe one input and one output module.

The 963's metering facilities are located on an overbridge at the top of the console. Studer has available for this console a group of "penthouse" modules, which include VU and PPM needle pointer meters; vertical bar graph meters with VU or PPM characteristics; limiter indicators; signaling, talkback, and intercom facilities; a multifunction digital clock; and remote controls for peripheral devices.

The switching at the top of the module (Figure 4–31) allows the operator to select one of eight submix groups. AUX controls 3 and 4 follow. The input module has a mixdown function for simultaneous changeover to the tape input of all input modules. The mixdown defeat button cancels this function. Below it is the overload peaker; the input level–audio generator rotary switch; the phantom power switch and phase-reversal switch; the mike presence-absence filter; shelving EQ controls for treble and bass and their in-out bypass button; AUX 1 and 2 controls; four master output-channel select buttons with panpot and pan switch; muting with LED, PFL, and solo buttons; and channel fader.

When the input selector is in the mike

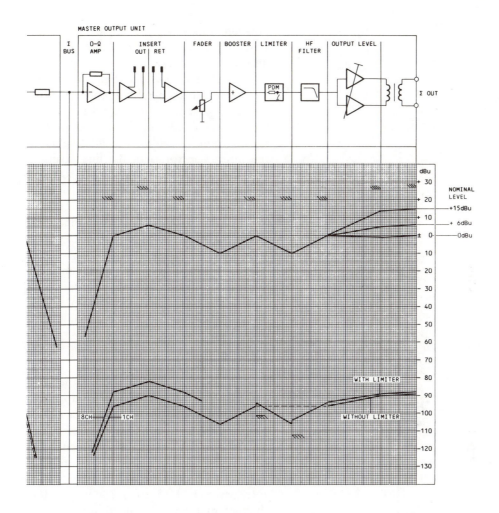

Figure 4–30   The Studer Revox 963 console. (Courtesy Studer Revox America, Inc.)

Figure 4–31 The 963 mono input module. (Courtesy Studer Revox America, Inc.)

position, a logic signal is applied to a logic signal buss as soon as the following conditions are satisfied: the channel fader is open, the master fader is open, the master selector has been actuated, and the mute button is inactive. This signal is used in the console's monitor module to control speaker muting and on-air signal lights. If the input selector is in the line position and the same conditions are met, logic signal is available for starting turntables or tape playbacks.

The input module is also available in stereo high level with or without equalization.

The two models of the master module (Figure 4–32) are virtually identical except for the top of the module strip, where the group master has the additional high-level input, switchable between external and group, and is used for mixing down the master output channel. Below are the limiter and compressor LED indicators and gain control; the limiter-compressor in switch; the link button to provide stereo coupling with the adjacent master channel; the limiter, switchable between master channel and group channel; release time control; ratio control, switch adjustable between 1.15 to 1.2 control; the channel high-level input rotary control, with PFL; AUX 1 and 2 output pots, selectable before (PF) or after (AF) the channel controller by a pot push-pull switch; the channel rotary gain control; selection switching with panpot; mute button with LED indicator; master channel PFL; and the master channel fader.

Other features for the Studer 963 console include the studio monitor and AUX master module, the control room monitor module, the monitor expansion module, tape machine remote-control module, and a patch panel that provides access to all input, master, and limiter-compressor insert points.

From the spec sheet:

1. Inputs: mike Z: 1.6 Kohm, line Z: 10 Kohm.

Figure 4–32 The 963 master module. (Courtesy Studer Revox America, Inc.)

2. Frequency response: 31.5 Hz to 16 KHz, +0.5, −1 dB.
3. Signal-to-noise ratio: >95 dB.
4. THD: <70 dB.

Rupert Neve Consoles

Rupert Neve Inc. is an international manufacturer of audio consoles with offices in the United States and the United Kingdom. Neve consoles have always been on the cutting edge of audio technology. We shall look at the Neve 51 series, the V series, Neve's NECAM 96 studio control system, and the Neve digital signal-processing system (DSP).

Neve 51 Series

The Neve 51 series consoles (Figure 4–33) are a range of consoles consisting of four distinct types, each available in differing configurations and numbers of input channels and groups. All are designed for use in either stereo or mono operation.

Type 5116 is available with 24, 36, or 48 input channels, including 24-track recording facilities and 24-track monitoring. It has eight subgroups of inputs and four main program outputs. All inputs and subgroups are provided with a four-band Formant spectrum equalizer (FSE), high- and low-pass filters, and limiter-compressor.

Type 5106 is identical to the 5116 but lacks a 24-track monitoring facility.

Type 5114 is designed with four subgroups and two main program outputs, using 12, 24, or 36 input channels. Two alternative types of input module are available. Type 5104 is similar to 5114 but includes an integral patch bay and either 16 or 24 input channels.

The reader can see that to describe the input-channel modules of a modular console pretty much describes the console itself. The 51 series input modules (Figure

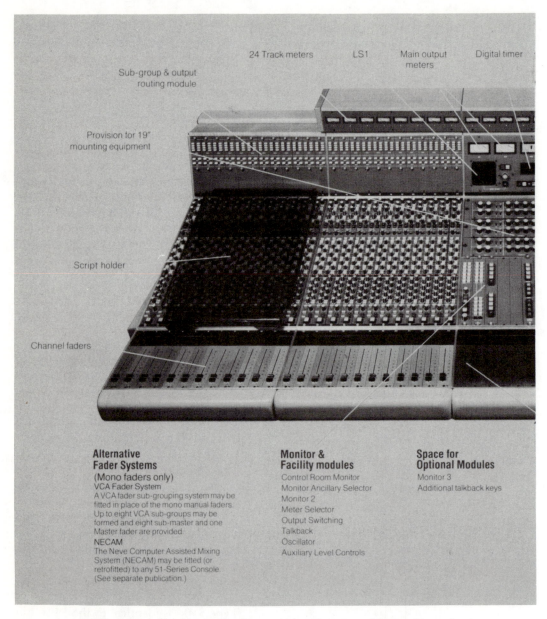

24 Track meters   LS1   Main output meters   Digital timer

Sub-group & output routing module

Provision for 19" mounting equipment

Script holder

Channel faders

Alternative Fader Systems
(Mono faders only)
VCA Fader System
A VCA fader sub-grouping system may be fitted in place of the mono manual faders. Up to eight VCA sub-groups may be formed and eight sub-master and one Master fader are provided.
NECAM
The Neve Computer Assisted Mixing System (NECAM) may be fitted (or retrofitted) to any 51-Series Console. (See separate publication.)

Monitor & Facility modules
Control Room Monitor
Monitor Ancillary Selector
Monitor 2
Meter Selector
Output Switching
Talkback
Oscillator
Auxiliary Level Controls

Space for Optional Modules
Monitor 3
Additional talkback keys

Figure 4–33 The Neve 51 series console. (Courtesy of Rupert Neve, Inc.)

4–34) are each constructed with a rigid steel frame carrying five separate self-contained subassemblies. Each is independently and easily removable for servicing or replacement.

The 51 series input modules include mono types 83010, 83049, and 83012.

The 83010 module is used on the 5116 and 5106 consoles. The 83049 and 83012 modules are used as alternatives on the 5104 and 5114 consoles. Type 83017 is the stereo input module.

The 83010 has an input section at the top of the module with mike or line avail-

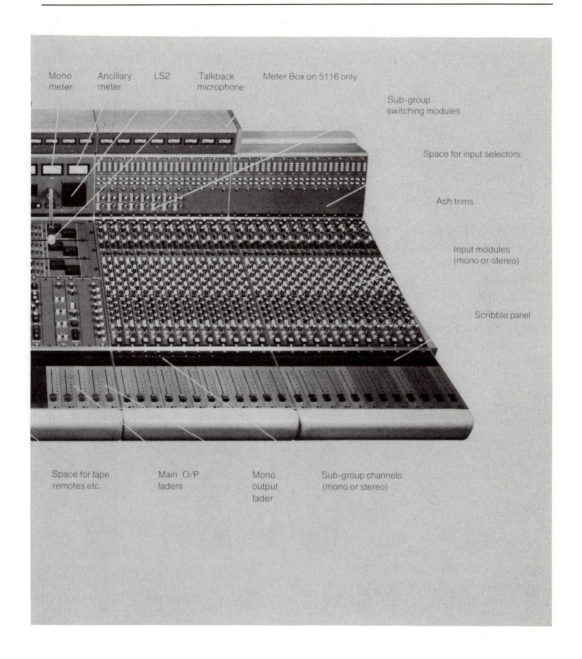

Mono meter · Ancillary meter · LS2 · Talkback microphone · Meter Box on 5116 only · Sub-group switching modules · Space for input selectors · Ash trims · Input modules (mono or stereo) · Scribble panel · Space for tape remotes etc. · Main O/P faders · Mono output fader · Sub-group channels (mono or stereo)

able, sensitivity switching, phantom mike powering, and phase reversal. Trim controls and a remote mike cut switch are also available. Below are the high- and low-pass filter and dynamics sections. Below the dynamics is the direct output section. It is at line level and may be used to feed a tape machine. The signal may be selected from either preequalizer or postequalizer, and a rotary gain control, PFL button, and on-off button with LED indicator are included in this section. Below the direct output are the two auxiliary output sections. Eight auxiliary outputs

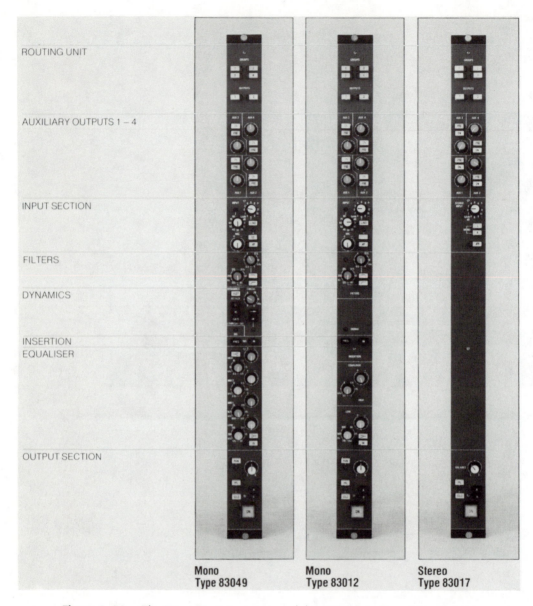

ROUTING UNIT

AUXILIARY OUTPUTS 1 – 4

INPUT SECTION

FILTERS

DYNAMICS

INSERTION
EQUALISER

OUTPUT SECTION

**Mono
Type 83049**

**Mono
Type 83012**

**Stereo
Type 83017**

Figure 4–34   The Neve 51 series input modules. (Courtesy Rupert Neve, Inc.)

are available on the 5116 and 5106 consoles, each with independent input-level controls, on-off switches, and prefader or postfader selection. Signal to AUX 1 through 4 is derived after the channel on switch, and on the 5116 console it follows the multitrack switching when the console is used in the multitrack mode, providing auxiliary outputs from the monitor mixdown section for use as reverb-on-monitor or cue-foldback feeds. Below the AUX outputs is a channel insertion section where a prefader insertion point may be selected before or after the equalizer. An insertion in-out button is provided. After the insertion section is a

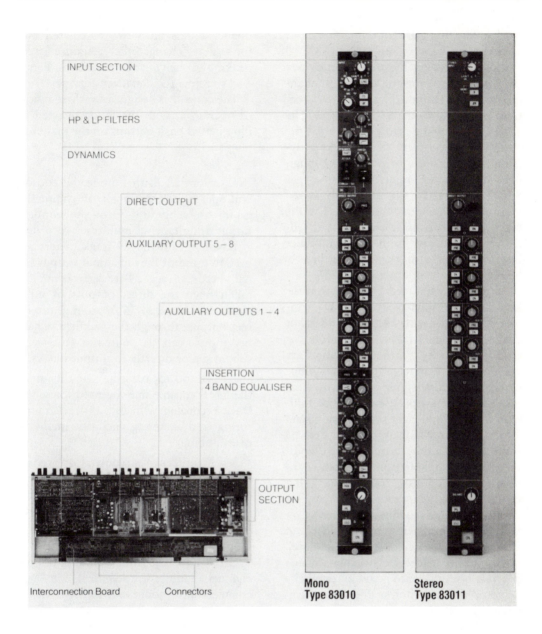

INPUT SECTION

HP & LP FILTERS

DYNAMICS

DIRECT OUTPUT

AUXILIARY OUTPUT 5 – 8

AUXILIARY OUTPUTS 1 – 4

INSERTION
4 BAND EQUALISER

OUTPUT
SECTION

Interconnection Board   Connectors

**Mono
Type 83010**

**Stereo
Type 83011**

four-band FSE equalizer and then the output section of the module. We will discuss the FSE equalizer later. Module output may be routed directly or through a stereo panpot to the subgroup or to the output busses through a separate channel routing module. The prefade listen button allows the PFL signal to be heard on the console speaker. A large illuminated button is used for channel on, and two master cut busses enable a group of inputs to be switched simultaneously through the master A and B buttons.

The 83049 input module has its routing switching at the top of the module, followed by the auxiliary outputs. This module has four AUX outputs. The input section is next and is similar to the 83010 module. Following the input section are the filters, the limiter or dynamics, channel insertion, equalizer, and output sections, all similar to the 83010 module. The 83012 input module closely resembles the 83049 module except that it lacks a dynamics section and has a two-band equalizer. The stereo input modules, types 83011 and 83017, are essentially stereo versions of the mono modules. On the Neve 51 series consoles, the vertical faders are not integral to the input modules but are mounted on the console directly below the input modules.

As stated earlier, the 51 series is available in four distinct types. The 5116 console is available with 24, 36, or 48 input channels, which can be either mono or stereo; 24-track recording facilities are available. There are eight mono subgroups of inputs or four stereo subgroups, four main program outputs, eight cue or foldback outputs, and eight reverberation send outputs. These 16 outputs are derived from eight auxiliary groups. There is direct output available from all input channels, stereo control room monitoring facilities, a secondary monitor system, bar graph or regular moving-coil metering on the main program outputs, prefader and afterfader listen, including "solo-in-place" facility, transmit-rehearsal switching, which controls both in-console and external circuits, talkback facilities, and a multifrequency line-up oscillator.

Formant Spectrum Equalizer

The Formant spectrum equalizer was designed to be related closely to both music and to the human ear. The four-band device has continuously variable frequency controls. The cut and boost controls are also continuously variable and have a range of ± 118 dB. The high- and low-fre-

quency bands have two associated buttons that change the response from peaking to shelving characteristics. Figure 4–35 graphs the equalizer curves.

The console is designed for three modes of operation, with the mode selected by illuminated push buttons on the main meter panel:

1. Normal. In this mode the console will operate conventionally in channel to group (or subgroup) to output configuration. Simultaneous multitrack recording is possible, and the 24-track meters are used to monitor the tape input-output signal. Feeds to the 24-track recorder are taken from the direct outputs of input channels 1 through 24 or from group direct outputs through the jackfield. Channel inputs can be routed to the eight subgroups or directly to output busses.

2. Multitrack recording. In this mode the 5116 console may be used for multitrack recording with full 24-track monitoring and metering facilities including overdubbing.

3. Multitrack mixdown or remix. Here, the 24-track output of a tape machine is routed directly to input channels 1 through 24, and full mixdown facilities, including subgrouping, are available.

Groups and subgroups. All mono input channel modules are identical, but eight channels have additional facilities to enable them to be used as groups or subgroups. If the channel is selected to be a group channel, the output may then be routed to the console output busses. If subgroup is selected, the signal can only be routed back to other group busses. In either case, the channel's input section may still be used as a mike- or line-level input to the console and may be routed to group or output busses, respectively, when the rest of the module is being used for group or subgroup. Rotary level controls and

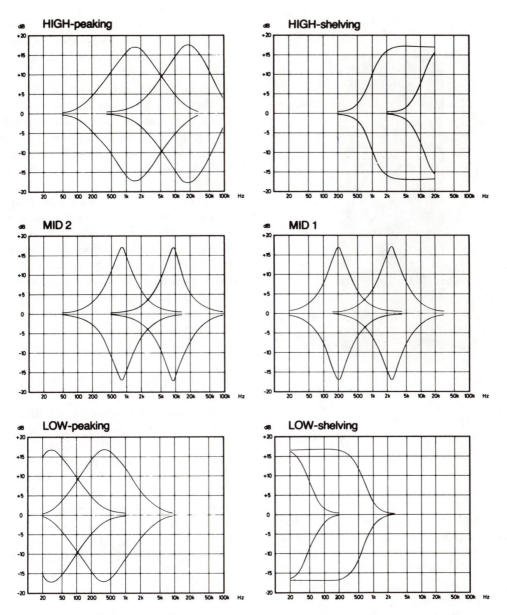

Figure 4–35   Neve FSE curves. (Courtesy Rupert Neve, Inc.)

panpots are provided, together with independent prefader and postfader listen push buttons. The equalizer and limiter-compressor may be selected to be either the group or the input circuit. Any number and combination of group-subgroups up to eight may be used, with or without the equalizer, and the full number of input channels can always be used. Figure 4–36 illustrates the group-subgroup module.

The signal from each input module is selected to subgroups or output busses through the routing buttons. On the 5116 and 5106 consoles these are on separate modules above the input channels. On the

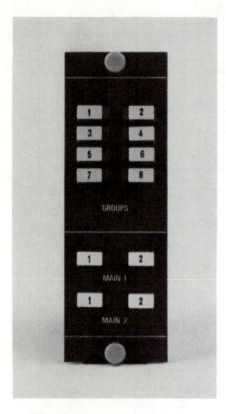

Figure 4–36   Neve channel to subgroup or output routing module. (Courtesy Rupert Neve, Inc.)

5104 and 5114 consoles the routing buttons are an integral part of the input module.

Console outputs. The main program outputs are individually controlled by master faders, and prefader monitoring is available on push switches appearing on the console speaker turret. The two line outputs on the 5116 and 5106 consoles normally derive their signal from outputs 1 and 2 but may be selected from outputs 3 and 4. The user can route tone or talkback to the line outputs. The mono output signal is derived from a mix of outputs 1 and 2 but may be selected from outputs 3 and 4 or from prefader or postfader output. The auxiliary outputs have master rotary-level controls and on buttons

with LED indicators. Independent outputs for reverb and cue-foldback ensure that talkback cannot be routed to outputs being used for reverberation.

Console loudspeakers. Two small integral speakers are mounted in the console, one for PFL and solo and one for return talkback to the console. Both have independent level controls and cut buttons and interlock with talkback keys to prevent feedback.

Oscillator. The oscillator may be routed directly to the console output lines at either a calibrated or variable level and at 40, 100, and 400 Hz as well as 1, 4, 10, and 15 KHz. The oscillator signal may be switched onto the 24-track tape machine through the direct outputs.

From the spec sheet (note how the Zs vary from American consoles):

1. Inputs: mike: Z In, 1.2 Kohm, line: Z In, 10 Kohm.
2. Outputs: nominal level: +4 dBu (0 VU), maximum output: +26 dBu into 600 ohm, output Z: 10 ohm.
3. Performance: noise: mike EIN: better than −125 dBu, 20 Hz to 20 KHz, terminated in 200 ohm. Line: better than −80 dBu, unweighted, 20 Hz to 20 KHz, distortion: Better than 0.03%, 40 Hz to 15 KHz, frequency response: ±0.5 dB, 20 Hz to 20 KHz.

V Series Consoles

The Neve V series consoles (Figure 4–37) are designed for use primarily in music recording. They are available in 36-, 48-, and 60-input channel configurations. The series V is built around the NECAM 96 studio control system, a digital system that is fully integrated into this analog console. It will be discussed separately. The series V has an independently assignable patch panel, allowing flexibility of inser-

Figure 4–37   Neve V series console. (Courtesy Rupert Neve, Inc.)

tion positions, a 200-segment bar graph metering device, and multiple modes of operation.

Figure 4–38 depicts an input-output module.

The top of the input section strip provides 48-track routing selection of signal sources through the 1 to 24 and 25 to 48 buttons. Panpotting is selected between odd and even busses using the pan switch to the right of the panpot. The bounce (B'NCE) switch simplifies track bouncing by sending the multitrack return back to the routing matrix by way of the monitor path. Below the routing section is the input section, which provides separate gain control of mike- and line-input levels on two continuously variable pots. Mike gain varies between +20 and +70 dB, and the line control provides ± 10 dB trim with a center detent at 0 dB. The mike control can disable phantom power on a per-channel basis if the user pulls the knob. The −30 dB mike pad switch below the mike pot allows line-level signal to be accepted by the mike pot. The C/O switch to the right of the line pot flips (reverses) the mike and line inputs, and a red LED indicates a "live" mike input. The phase button beneath the line pot phase reverses the mike- and line-input signals, and the group (GRP) button below it provides patch-free subgrouping of the input channel.

The filter section has a high- and a low-pass filter, with 12 dB per octave slopes

Figure 4–38   Neve V routing, input, and filters. (Courtesy Rupert Neve, Inc.)

at frequencies that vary continuously from 31.5 to 315 Hz and 7.5 to 18 KHz. The filters are switched in and out when the operator pulls the appropriate knob. A yellow LED is illuminated when either filter is in the circuit.

The dynamics unit (Figure 4–39) is a combined limiter-compressor and gate expander. The gate-expander controls, on the left of the module section, are switched into the circuit with the gate switch at the lower left. The limiter-compressor controls on the right are activated by the limiter-compressor switch at the upper right. The gate-expander controls are KEY, a switch that when depressed allows the gate expander to be triggered by an external source inserted into the dedicated key input on the patchbay; INV, which inverts the sense of the gate-expander control; and HYST, which adjusts the amount of gate hysteresis (the dB difference between the gate mute level, set by threshold, and its unmute level); THR, threshold; RGE, range or depth of the gate expander; and REL, gate-expander release time.

The limiter-compressor controls are right arrow button, which links limiter-

compressor control voltage to the next input channel to its right, for stereo operation; GAIN, which provides independent control of gain makeup over a 30 dB range; THR, threshold adjustment; RAT, which sets compression ratio; and REL, which sets the compressor normal release time. The dynamics unit is metered with a tricolored LED mounted in the meter section of the console, which indicates gain reduction.

Below the dynamics unit are the auxiliary sends (Figure 4–40). They get signal source from either the channel itself or from the monitor path assigned from the switch matrix to the right of the vertical fader, farther down the channel strip. They may be configured as either eight mono sends, with individual on-off switching

Figure 4–40 Neve V auxiliary sends. (Courtesy Rupert Neve, Inc.)

Figure 4–39 Neve V dynamics unit. (Courtesy Rupert Neve, Inc.)

through the on button, or as stereo sends by switching the ST button. The odd-numbered level controls (1, 3, 5, and 7), act as panpots in ST. PRE switches the source point for the sends, prefader or postfader, in pairs.

Operationally, the PRE-AUX buttons send signal to performers in a studio in the console tracklaying mode and also provide some effects sends in mixdown mode.

Below the AUX sends is the Neve FSE equalizer (Figure 4–41), which was mentioned when we discussed the 51 series consoles. Here controls are provided to switch the equalizer: EQ activates the device; DYN connects the equalizer into the dynamics unit; INS positions to either the channel or monitor path; and PREQ is a configuration switch.

The bottom of the channel strip has the track level control pot (Figure 4–42). This control is center detented at line-up level,

with 10 dB of gain available. The DIR switch sends the channel signal directly to its corresponding recorder track. The small fader, together with its solo and cut buttons, is assigned either in the channel or in the monitor signal path by master status or is switched individually by the swap button. Swapping the faders does not reassign the auxiliaries. C/O reverses

Figure 4–42   Neve V fader controls. (Courtesy Rupert Neve, Inc.)

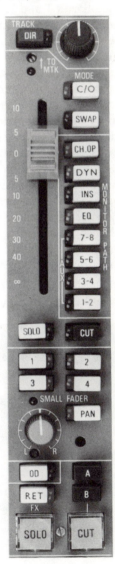

Figure 4–41   Neve V FSE equalizer. (Courtesy Rupert Neve, Inc.)

the master console status locally, for record and mixdown, and can set overdubs and submixes routed through the multitrack matrix during mixdown. The LED above the small fader indicates when this fader is controlling the signal that is feeding the multitrack matrix. The small fader cut button is electronically switched and may be automated with the NECAM 96 events controller.

The monitor path selectors to the right of the fader assign the channel strip dynamics (DYN), insertion (INS), equalizer (EQ), and auxiliaries 1 and 2, 3 and 4, 5 and 6, 7 and 8, into the monitor signal path. Ch.op connects the channel postfader output to monitor path input and can allow additional effects sends to be set up through the multitrack matrix. These selector buttons are grouped for easy viewing so that each module's signal structure can be quickly assessed.

The output section allows the operator to select the busses for mixdown or for simultaneous monitoring during recording. Panning is selected between left-right (odd-even) busses. The small fader LED illuminates when it is feeding output in this section. The overdubbing switch below the panpot provides correct monitor-

ing and cues when overdubbing in conjunction with master monitor status. The monitor may be switched without affecting the cue sends. The effects return button individually "solo safes" both signal paths.

Below each module is the channel's large vertical fader. Its path assignment is controlled from the master status and the swap button above. The red A and B status buttons allow groups of large faders to be cut simultaneously by the master controls. This function may be automated using the NECAM 96 events controller. Figure 4–43 shows the input-output module removed from the console, lying on its side.

The Neve V console's central facilities are shown in Figure 4–44, with several input-output modules on either side. The central facilities are divided into upper and lower parts, and the lower part is divided into four quadrants. The upper right quadrant has the return talkback speaker, push-to-talk button, and gain control. Below it is the NECAM 96. The upper left quadrant has the studio monitor selectors.

Below the monitor section are the console status configuration controls. Using

Figure 4–43   Neve V input-output module. (Courtesy Rupert Neve, Inc.)

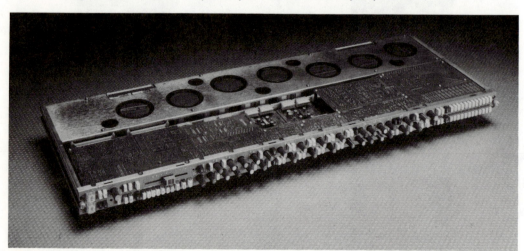

Figure 4—44   Neve V central facilities. (Courtesy Rupert Neve, Inc.)

the master mode selector buttons, MIC, FADER SWAP, MIXDOWN, and BROADCAST, the console can be configured so that in tracklaying mode the channel path, normally in mike input, is sent to the multitrack recorder through the routing matrix. Monitoring is achieved by mixing to the main two track (2T) outputs through the monitor path. In tracklaying mode, the prefade auxiliaries are taken to precut to achieve in-place soloing without cutting cue sends. In mixdown mode the channel-path line inputs are routed to the 2T outputs to handle multitrack replay for the output mix. In mixdown mode the prefade auxiliaries are taken postcut to facilitate use for effects sends. The monitor path is routed to the multitrack matrix and can be used for additional mix inputs, effects returns, and effects sends, which are sourced from the channel postfade output. Broadcast mode provides "simulcast" mixing, allowing simultaneous multitrack and broadcast production work.

The Neve V console allows for master, mike and line switching to complete the status configuration. These master flip functions can operate on the whole console or can be split on either side of the central section.

The V series solo system combines the monitoring facilities of prefade listen (PFL), afterpan positional listen (APL), and solo-in-place, or cut solo. All console solo buttons are electronically latched, providing individual but identical solo facilities for both signal paths regardless of fader swap. The CHAN SAFE and MON SAFE buttons provide individual path solo-safe controls that can be linked to tape machine record functions or "on-air" signaling for automatic changeover.

On the central facilities panel, above the four quadrants and on the left, is the signal indicator master threshold control. It adjusts the level at which the indicator LED above each multitrack meter signals. Below it is a calibrated multifrequency oscillator. To its right are the AUX masters. Their controls may be configured between eight mono and four stereo outputs, switchable in pairs. These outputs are taken through the patchfield, after which they split off to become foldback with talkback and effects sends with no talkback. To the right of the AUX masters are two cue mix sections and four stereo REV returns. The two stereo cue sends can be sourced from the auxiliaries, the control room monitor, the patchbay, and 2T outputs and may be blended to provide the total cue signal. Four stereo REV return units may be fitted to the console.

Metering for the series V console is provided above each input-output module with a 200-segment plasma vertical bar graph. VU and PPM characteristics are available simultaneously. Eight auxiliary master VU meters are also provided.

From the spec sheet:

1. Inputs: mike: Z >1 Kohm, balanced, line: >10 Kohm.
2. Outputs: track, Z less than 15 ohm.
3. Performance: THD: better than 0.04%, 20 Hz to 20 KHz, frequency response: flat $+0.5$ dB, -1 dB from 20 Hz to 20 KHz.

The Neve NECAM 96 Studio Control System

The Neve NECAM 96 is a system for computer assistance in console operations too complex for the operator to handle by herself but which require lightninglike response.

The NECAM 96 panel control unit, located centrally on the series V console, is shown in Figure 4–45.

Stand-alone NECAM 96s have three user-associated parts, three rack-mounted parts, and two peripherals, one in the console and one in an associated tape recorder. The user parts are the control unit, which is a laptop computer keyboard with keys similar to those shown in Figure 4–45. There are some dedicated keys and some "smart" keys; the rest is a typewriter input keyboard. The second user part is the display unit, essentially a 14-inch color television RGB screen. The third user part is the special faders on the console, which are servomotor driven as well as finger operated. The NECAM 96 can control up to 96 faders.

Mounted in a rack nearby are the computer unit, the floppy disk drive, and the power supply. The analog to digital and digital to analog converters are mounted within the console or its base frame. The intelligent tape peripheral is mounted in association with the tape machine.

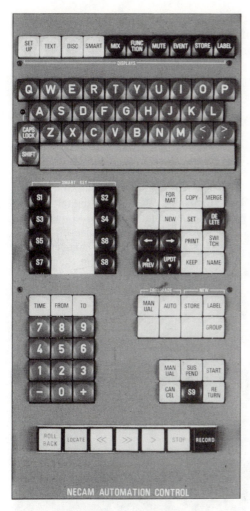

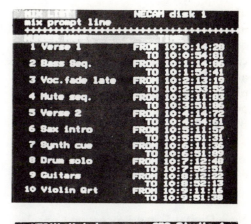

Figure 4–46 Neve NECAM displays. (Courtesy Rupert Neve, Inc.)

Figure 4–45 Neve NECAM 96 control panel. (Courtesy Rupert Neve, Inc.)

The heart of the system is the computer, with its disk drive. The drive can store and mix data on disk. It can control 128 "events," which the operator enters, in plain English, from the keyboard, including adjustable tape preroll, small fader mute, equalization in and out, insert, and effects in and out. There can be mix and event merging, with user selection of tracks and functions to be merged. When the faders are manually moved, computer update is automatically done. User-pro-

grammable smart keys reduce repetitive instructions of up to 60 keystrokes to a single key push. Computer program users will know this as using macros. The display, on top of the console, provides instant system status, such as the last label passed and the one coming up. There is a comprehensive list and display of information, such as labels, mutes, stores, events, mix, smart keys, text pages, setup, and main display, with a combination of mix, labels, stores, events, and time-code information. Figure 4–46 shows the NECAM 96 displays.

Up to 99 fader settings can be stored for instant callup. The operator can crossfade to a stored fade or merge point, using programmable crossfade time or manual fader control.

NECAM 96 uses SMPTE/EBU time code because it interfaces with both audio and video tape recorders. It can use bi-phase, another time code, for timing and positional input, when it interfaces with film machines.

There can be off-line mute and events editing, and system setup and diagnostic programs are available.

Neve Digital Signal Processing Console

The Rupert Neve company, in its descriptive literature about the digital-signal-processing (DSP) console, expresses its

concern about the back and forth transfer of audio, from digital to analog to digital, as follows:

> Once audio is in digital form—encoded in a stream of numbers—its quality is "sealed in," and the signal is very robust indeed as long as it remains digital. However, most processes in the sound studio involve transfers from one tape to another for track-bouncing, mixdown, post-production editing, or changes in level, equalization, echo and dynamics. Without a digital mixing console, the signal must be reconverted to analog form, losing the sealed-in quality, even for the most minor changes. Such processes occur many times in the production of the average recording and inevitably there is a loss of audio information at every

Figure 4–47   The Neve DSP console in close-up. (Courtesy Rupert Neve, Inc.)

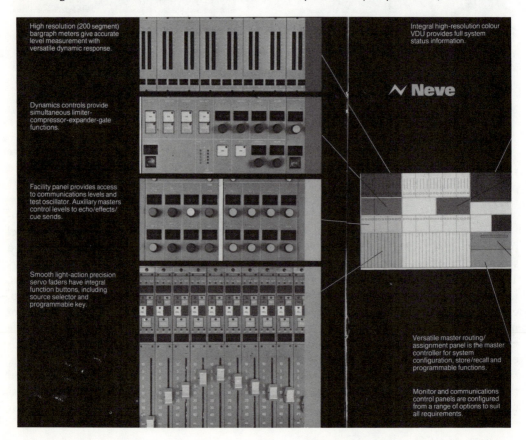

High resolution (200 segment) bargraph meters give accurate level measurement with versatile dynamic response.

Integral high-resolution colour VDU provides full system status information.

Dynamics controls provide simultaneous limiter-compressor-expander-gate functions.

Facility panel provides access to communications levels and test oscillator. Auxiliary masters control levels to echo/effects/cue sends.

Smooth light-action precision servo faders have integral function buttons, including source selector and programmable key.

Versatile master routing/assignment panel is the master controller for system configuration, store/recall and programmable functions.

Monitor and communications control panels are configured from a range of options to suit all requirements.

conversion. The result can be that the final CD or broadcast fails to realize its full musical quality potential.

Neve thus created a digital-signal-processing console that benefits the quality of the product by eliminating the intermediate analog and digital conversions.

The first look shows that the DSP console (Figure 4–47) is smaller than most consoles with so many facilities available to the operator. The signal path can be configured as the operator requires. Fader layout can be switched so that the most important controls are the closest to the operator's hands. The left-handed operator is comfortable with the controls. The input units may be in another room or building, with only their controls in front of the operator. In fact, the input units can be up to a half mile away from the console, coupled by fiber-optic links. Each link carries 16 audio channels and four communication channels, so the number of lines is minimal. The DSP console can be as big as needed. A 56-channel console is accommodated within two arms' span. Faders can be organized in musical order regardless of where the mikes are plugged in. For stereo operation the ganging of input, subgroup, and output paths, with two-channel processing operated from one set of controls, provides a compact working layout.

The details of console setup, EQ, dynamics, system routing, and the setting of

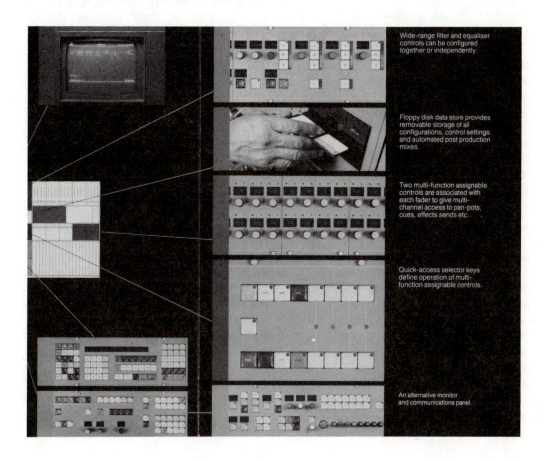

Wide-range filter and equaliser controls can be configured together or independently.

Floppy disk data store provides removable storage of all configurations, control settings and automated post production mixes.

Two multi-function assignable controls are associated with each fader to give multi-channel access to pan-pots, cues, effects sends etc.

Quick-access selector keys define operation of multi-function assignable controls.

An alternative monitor and communications panel.

every control can be stored on computer disk for instant recall and reset. Previous recording session settings are restored at will, so session setup time is eliminated. The producer can take away a floppy disk with special routing and configurations to be used on a future program.

Every channel and group has a variable-signal time-delay feature available, which can correct for mike positioning. This feature opens up the possibility of equalizing the arrival time of signals from the mikes at different distances from the same source, retiming synthesizer tracks, or, in combination with mixing and EQ, creating phasing and reverb or other, as-yet-undiscovered effects. Up to 2.5 seconds of delay is available for each 24-channel processing unit, which can be spread as required in any or all of the units.

The DSP console employs, in addition to the typical Neve bar graph meters, a video display (VDU) which shows the position of any signal processor, and its arrangement in the signal path, in simple schematic form.

Provision is made for analog interfaces at the inputs, outputs, and insertion points, eliminating the need for a patch panel.

Operation of the DSP console starts with the internal computer's start-up menu, which lets the operator ask the system to check itself with a diagnostic program or go directly to operation. The diagnostic software also runs during operation so that processing is checked continuously. When the operator selects the operation mode, a range of options is presented in color on the VDU. The console will (1) continue exactly as it was set up when last turned off, (2) continue as last time but with all controls zeroed, ready for entirely new settings, (3) continue with the configuration as last time, but with a new set of control settings recalled from floppy disk, or (4) present new system

configurations selected from four user-specified options embedded in console memory or something selected from disk. These options may include tracklaying, multitrack mixdown with overdub option, direct stereo mix with clean feeds, and special dubs with multiple equalized echoes and flanges.

Routing from any inputs to any outputs may be checked on the VDU and changed if necessary on the assignment panel by entering, for example, "input 32, route to group 13, execute." Alternately, the same input and group can be called up from their fader position by pressing the relevant access buttons.

Audio balance is set as usual on the input and group faders. Pressing the ACCS button above each fader readies the set of processing controls. Stereo signals arriving at the inputs are handled by defining two inputs and their relevant mixing groups as a stereo path; then all controls and processors are ganged and work from a single set of controls. If these settings have been previously memorized, or saved on disk, they can be recalled. Echo and effects insertion and return points, from either analog or digital facilities external to the DSP, are assignable anywhere within the console. Talkback, cue and foldback, and even PA are available from the console, as are the normal mute features.

We noted earlier that the console itself is small, but it is served by input racks that accept up to 16 mike or line inputs each. Here mike signal is analog amplified, and both mike and line signal are converted to digital signal. The digital signal is then multiplexed on a fiber-optic link to the main processor rack. The processor racks are contained in three six-foot cabinets, the heart of the system, and are remote controlled from the console by digital instructions, also linked by fiber-optic cables. Neither the audio signal nor its digital equivalents pass through the console itself. The console is only an array of controls with associated graphic displays, keeping it within human operating dimensions. Figure 4–48 presents a block diagram of the DSP console.

From the spec sheet:

1. Input Z, mike or line, 10 Kohm.
2. Output Z, 200 ohm balanced, analog.
3. A/D & D/A system performance: sampling rate: 48 kHz, quantization: 16 bits, all inputs optimally dithered, signal-to-noise ratio: >89 dB, frequency response: +0.2 dB to −0.5 dB, 20 Hz to 20 kHz, THD at +22 dB level, 20 Hz to 20 kHz, <0.05%.

TASCAM Consoles

TASCAM is the professional products division of TEAC. We will look at two of its products, which are at opposite ends of the console input spectrum in terms of number of inputs.

TASCAM M208

The TASCAM M208 (Figures 4–49 and 4–50) is a transportable eight-channel mixing board. There is also a model M216, a 16-input channel board. All input and output connections to this console are made on XLR-3, RCA pin jack, or TRS connectors on the back of the console.

The eight-channel strips each have the following features, from top to bottom of the strip: a 30 dB pad in-out button for mike and line; a 44 dB rotary trim pot; a tape selector that selects either the tape or the mike-line input on the console back; four equalizer controls; overload LED; foldback level control; tape selector switch; effects level control; two-program-channel assign buttons for four pro-

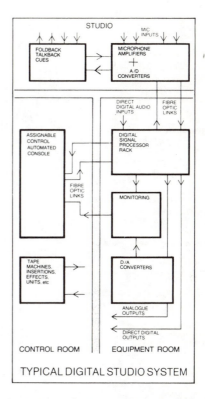

Figure 4–48   The Neve DSP typical block diagram. (Courtesy Rupert Neve, Inc.)

gram channels; panpot; prefader listen switch; and input channel fader.

There are five submix systems: main mix, stereo mix, foldback, effects, and solo. There are four program busses, each with a master fader and pan control, with a choice of either XLR-3 or RCA pin jack output. The master faders are located under the four VU meters. Meter switching places the meters on the required output busses, and each VU meter has an LED peaker. The effects submix buss permits the input of echo, delay, flanging, and reverb by taking signal from the channels, sending that signal to external points, and returning the effects-altered signal through individual pan and gain controls to the stereo buss. The stereo left and right master faders are to the right of the program master faders.

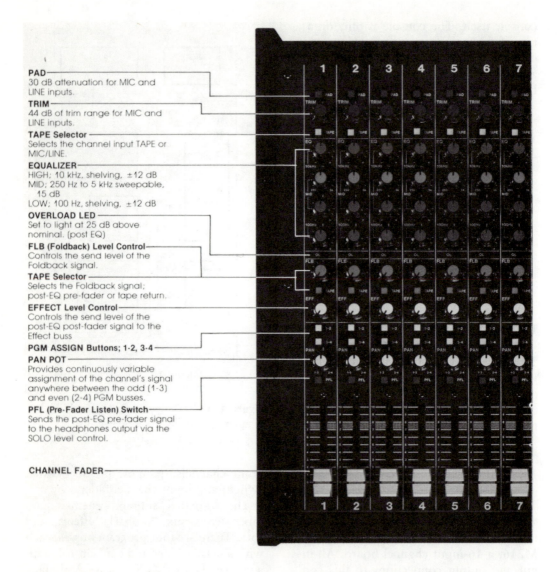

PAD
30 dB attenuation for MIC and
LINE inputs.

TRIM
44 dB of trim range for MIC and
LINE inputs.

TAPE Selector
Selects the channel input TAPE or
MIC/LINE.

EQUALIZER
HIGH; 10 kHz, shelving, ±12 dB
MID; 250 Hz to 5 kHz sweepable,
    15 dB
LOW; 100 Hz, shelving, ±12 dB

OVERLOAD LED
Set to light at 25 dB above
nominal. (post EQ)

FLB (Foldback) Level Control
Controls the send level of the
Foldback signal.

TAPE Selector
Selects the Foldback signal;
post-EQ pre-fader or tape return.

EFFECT Level Control
Controls the send level of the
post-EQ post-fader signal to the
Effect buss

PGM ASSIGN Buttons; 1-2, 3-4

PAN POT
Provides continuously variable
assignment of the channel's signal
anywhere between the odd (1-3)
and even (2-4) PGM busses.

PFL (Pre-Fader Listen) Switch
Sends the post-EQ pre-fader signal
to the headphones output via the
SOLO level control.

CHANNEL FADER

Figure 4–49 TASCAM M208 console. (Reprinted with permission from TEAC Corporation of America.)

From the spec sheet:

1. Input: mike Z: 2 or 8 Kohm, line Z: 22 Kohm.
2. Output: Z: 100 ohm.
3. Frequency response: any input to any output, 20 Hz to 25 KHz, +1 to −2 dB.
4. Signal-to-noise ratio: >57 dB.

5. THD: mike in to program out: 0.03%, line in to program out: 0.025%.

TASCAM M600 Series Consoles

The M600 consoles (Figure 4–51) are available in 24- and 32-input models and are designed for the professional record-

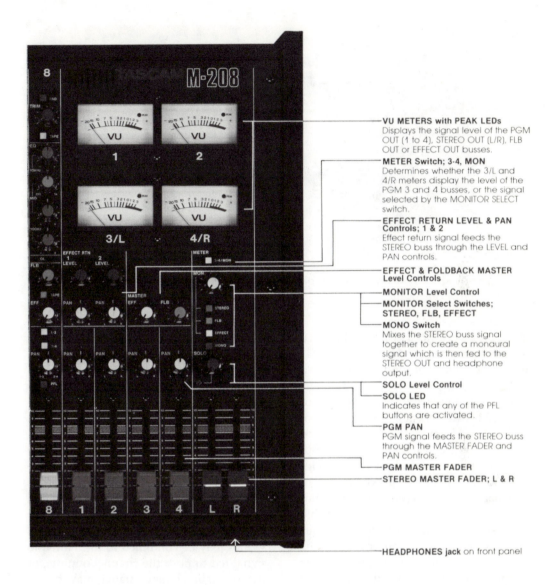

VU METERS with PEAK LEDs
Displays the signal level of the PGM OUT (1 to 4), STEREO OUT (L/R), FLB OUT or EFFECT OUT busses.

METER Switch; 3-4, MON
Determines whether the 3/L and 4/R meters display the level of the PGM 3 and 4 busses, or the signal selected by the MONITOR SELECT switch.

EFFECT RETURN LEVEL & PAN Controls; 1 & 2
Effect return signal feeds the STEREO buss through the LEVEL and PAN controls.

EFFECT & FOLDBACK MASTER Level Controls

MONITOR Level Control

MONITOR Select Switches; STEREO, FLB, EFFECT

MONO Switch
Mixes the STEREO buss signal together to create a monaural signal which is then fed to the STEREO OUT and headphone output.

SOLO Level Control

SOLO LED
Indicates that any of the PFL buttons are activated.

PGM PAN
PGM signal feeds the STEREO buss through the MASTER FADER and PAN controls.

PGM MASTER FADER

STEREO MASTER FADER; L & R

HEADPHONES jack on front panel

Figure 4–50   TASCAM M208, rear view. (Reprinted with permission from TEAC Corporation of America.)

Figure 4–51 TASCAM M600 console. (Reprinted with permission from TEAC Corporation of America.)

ing environment. Both consoles have 16 program busses to feed 16-track tape machines. Two monitor options include a choice of 16 monitor returns or, alternately, 32 monitor returns for use with 24- or 32-track machines. Both mono, which TASCAM calls standard, or stereo input sections are available for the M600.

At the top of the standard section (Figure 4–52) is a phantom power switch; a 30 dB pad switch; independent rotary trim controls for mike and line; a switchable high-pass filter; an EQ section with in-out switch; a phase-inversion switch, four AUX send controls, switchable to cover the eight AUX busses; and eight program buss assign switches, based on an odd-even channel system, with the pan control functioning between the selected buss pairs. The channel signal can also be assigned to, and panned across, the left-right stereo buss in the same way, with a channel on-off switch permitting the entire channel to be muted or engaged as required, a solo switch for monitoring the individual channel or groups of channels, and an overload LED indicator. The channel fader is mounted separately from the channel strip to allow easy conversion to VCA faders.

The stereo input (Figure 4–53) handles stereo sources such as compact disks (CDs), turntables, or two-track tape or the return of stereo effect signals. Dual stereo-line inputs are selected by a switch at the top of the strip. This is followed by a phase-inversion switch on the right channel and a trim control for matching the channel's input sensitivity with the source output level. The switch group below this allows the stereo signal to be configured as follows: left and right channel signals may be reversed, left and right channel signals may be combined to form a mono signal fed to both the channel outputs, or left or right channel signal may be selected and fed to both channel outputs.

A three-band EQ system follows, with in-out switch, an assignment switch matrix that assigns to odd-even program busses like the mono strip, as well as to the left-right stereo buss, then a balance control between the left and right channel signals, a channel on-off switch, a solo switch, an overload indicator, and the separately mounted channel fader.

The monitor strips, which are found in the monitor and program section of the M600, are available in single and dual monitor strips. The single monitor ver-

Figure 4–52   Standard input section. (Reprinted with permission from TEAC Corporation of America.)

Figure 4–53   Stereo input section. (Reprinted with permission from TEAC Corporation of America.)

sion (Figure 4–54A) has the 16-monitor return channels positioned directly above the 16-buss master faders. The monitor channels have essentially the same control layout as the input channel strips. Input to a monitor channel can be derived directly from its corresponding master program buss or from the tape return input by pressing the return switch at the top of the strip. Next follows a phase-inversion switch and a trim control. The monitor EQ section is the same as for the input channels, with an additional program EQ switch, which allows the EQ to be inserted into either the monitor channel or the corresponding program buss. Monitor channels have the same AUX send control configuration as the input channels. A fader is built into the monitor strip, but its operation can be exchanged with the master buss fader below it by pressing the fader reverse (REV) switch. This affords greater monitor level setting flexibility and permits direct mixdown using the monitor channels. The SUB (subgroup) switch sends the master program buss signal to the console's stereo buss. This allows odd-even pairing of program busses to form stereo subgroups. The monitor channel pan control pans the monitor signal across the master stereo buss. The channel on-off switch allows the monitor channel to be muted or engaged, and the solo switch allows individual channel monitoring.

In the dual monitor version (Figure 4–54B), each of the 16 monitor strips above the master program buss faders contains two monitor channels, which accept any line-level signal, making them useful as additional effects returns. A meter switch on top of each strip determines whether the upper or lower group of monitor channel signals are fed to the console's 16 VU meters.

The master section (Figure 4–55) is a wide strip with several functions. The strip is located between the input sections and monitor-program sections for easy access and operation. The two leftmost subsections of the master section include the eight AUX buss master level controls; below them are the effects return controls. Two effects return sections are provided, with their levels controlled by the faders directly below the strips. The effects returns are assigned and panned odd-even by switches and a rotary control. Two AUX send controls, on-off return switches, and solo switches complete the strips.

The control room strip, to the right of the AUX masters, incorporates the control room monitor assignment and level controls. A group of assign switches sends any of the eight AUX buss signals, the signals from a choice of three two-track reproducer tape inputs, or the master stereo buss signals to the control room monitor outputs. Two pairs of control room monitor outputs are provided and can be independently switched on-off using the A and B switches. The rotary control room control sets the control room monitor level. The control room strip also includes the master solo and PFL controls. A dim switch reduces monitor output level by 30 dB to permit conversation or phone calls. A mono switch combines the left and right channel control room monitor signals to form a mono signal used for phasing checks.

The final strip in the master section, the meter-studio-talkback strip, includes a group of meter assign switches that determine whether the console's left and right VU meters show control room monitor levels, studio monitor levels, or levels on the master stereo buss.

The studio monitor control section permits assignment of the stereo master buss or control room monitor signals to the studio monitor feed. The studio monitor signal can be summed to mono, and a master studio level control sets the monitor level. A phones-level control and earphones jack are provided There is also a

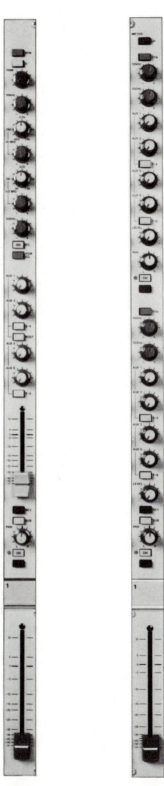

Figure 4–54 Single *(left)* and dual *(right)* monitor sections. (Reprinted with permission from TEAC Corporation of America.)

Figure 4–55 Master section. (Reprinted with permission from TEAC Corporation of America.)

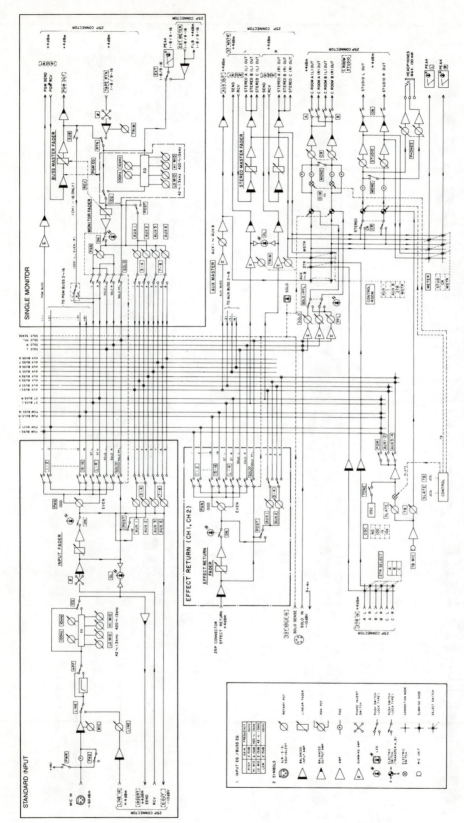

Figure 4–56 Block diagram. (Reprinted with permission from TEAC Corporation of America.)

90

master trim control for fine adjustment of master stereo output level.

The talkback section has a built-in mike, multitone oscillator, slate and talkback level controls, and assign switching to the 16 master program busses and four AUX busses.

Figure 4–56 is a truncated functional diagram showing the standard input, single monitor, effects return, and AUX master buss.

From the spec sheet:

1. Inputs: mike Z: 2.2 Kohm, line Z: 40 Kohm.
2. Outputs: 100 ohm.
3. Signal-to-noise ratio: >53 dB.
4. THD: <0.015%.
5. IMD: <0.03%.
6. Frequency response: mike to pro-gram out: +0.5 dB, −1.5 dB, line in to program out: +0.5 dB, −1.5 dB.

REVIEW QUESTIONS

1. What is a sisterboard? A motherboard?
2. What is a mainframe? A module?
3. What do we find in a spec sheet?
4. Explain input module buss assignment.
5. Describe some of the features that many professional consoles have in common.
6. What are some of the features unique to some professional consoles?
7. What is NECAM 96, and why is it used?

5

CONSOLE INPUT FACILITIES

In our discussion of professional control boards, we alluded to console inputs without describing them fully. Console inputs are outputs of other devices that feed their signal into a console. Examples are microphones, lines, tape playbacks, cart machines, and turntables. They are described as being either *low level* or *high level*. Microphones, whose output level is −50 to −60 VU, are low-level inputs. This indicates that their minute audio signal must enter the console through an input channel that has a preamplifier to boost the signal. Tape playbacks and external audio lines are examples of high-level inputs, often having outputs in the +4 to +8 VU range. Their signal enters the console through an input that either has no preamp or can bypass the preamp. A turntable can be high or low level depending on whether a preamplifier has been built into it.

MICROPHONES

Mikes can be discussed or described in many ways. Here we will discuss them by type, and their specific use is described in Chapter 11. First, though, a disclaimer: Modern professional mikes are, in general, very hardy instruments. One com-

pany representative used to display his product by hammering nails into a board with a mike and then proceeded to demonstrate that the mike still worked well. Broadcast and recording company executives who control the purse strings are loathe to discard anything technical or replace anything that still works. Thus most broadcast facilities still use microphones that are 15 to 20 years old. This text, then, would do a disservice to its readers if it did not mention, describe, and illustrate equipment that is found in most audio operations as well as state-of-the-art devices.

Mikes can be described by their directionality, since all mikes fall into one of two categories of directionality: directional or nondirectional in sound pick-up. Examples of directional mikes are the figure eight and the cardioid, which are discussed later. An example of the nondirectional mike is the dynamic. Some other ways of saying nondirectional are omnidirectional or all-directional. Some mikes are switchable to either category.

Mikes can also be described by the specific component of the sound wave that actuates them. Thus, dynamic mikes that employ either a metallic or plastic diaphragm in conjunction with a moving coil, or condenser mikes with a capacitive actuating element, are operated by the wave's pressure component. Ribbon mikes, which employ a small corrugated or washboard-shaped aluminum element, are actuated by the wave's velocity component.

Mikes can be described by their impedance (Z). The Z of a mike is the AC resistance that it presents at its output. All professional mikes have a low Z (150 to 200 ohm) and are balanced to ground. They have three output connector pins. Recall that in our discussion of consoles, the spec sheet described the required input mike impedance; all professional consoles require low-impedance mikes.

A specific method of describing mikes' directionality is by their polar patterns. A microphone's polar pattern is a polar (circular) graph description of the direction(s) in which the mike will either pick up or reject sound in relation to a graph axis at the center of the actuating element.

Thus, the dynamic mike has a circular pattern, the ribbon mike has a figure-eight bidirectional pattern, and the cardioid mike has a unidirectional heart-shaped pattern. Variations of the cardioid mike, the supercardioid and hypercardioid, narrow the heart-shaped pattern. Older cardioid mikes used a combined diaphragm and ribbon element to achieve the unidirectional pattern, or either of the other two pick-up patterns, with a recessed screwdriver-operated switch in the body of the mike. The newer cardioid mikes, and their variations, use a diaphragm-actuating element and the effects of sound phasing, using slits or holes in the body of the mike, behind the actuating element to achieve selective sound cancellation and thus achieve the unidirectional polar pattern. Mike polar patterns are discussed in Chapter 11. Finally, microphones can be described by reading their frequency-response curves. These too are discussed in Chapter 11.

Connectors

Microphones are physically as well as electronically connected to a studio's control console through a connector on the end of the mike, and the mike cables (Figure 5–1) are connected to connector receptacles on the studio walls (Figure 5–2). These connectors, called XLRs (or XLR-3s if they have three pins or receptacles), are standard in professional audio, if indeed there are any connector standards. They were introduced by Cannon years ago to replace the company's much larger P-type connectors, still found in ancient installations. The XLR-3 connector is now made by Switchcraft, as its A3 series, and by Neutrik of Switzerland.

Similar to other connector pairs that "mate" to complete a circuit, XLR-3s come in male and female types. It is standard in the business for mikes to terminate in a male connector. It follows, then, that if the mike has a male connector, its output cable must have a female connector at the mike end of the cable, and a male connector at the other end to mate with either another (extension) cable or a wall receptacle or equipment input. Wall receptacles and equipment inputs always use a female connector, while equipment outputs always use a male connector.

The XLR-3 connector used for audio work, whether male or female, has either three or four solder connections at its back end and three pins or three receptors at its front or "business" end. Other XLR connectors with more than three terminations are used for other circuitry purposes. In the audio XLR-3, these terminations, or pins, are labeled pin number 1, 2, and 3, with the shell being the connector's metal enclosure. Pin number 1 is usually signal ground (earth), and in the mike cable and extensions it is also connected to the shell and to the cable shield. This is necessary to carry the outer

A 3 F

A 3 M

B 3 M

C 3 M

Figure 5–1 Mike cable connectors. (Courtesy Switchcraft, Inc., a Raytheon Company.)

Figure 5–2 Mike wall receptacles. (Courtesy Switchcraft, Inc., a Raytheon Company.)

case of the mike eventually to ground. The practice of connecting pin number 1, at the equipment end of the cable, to the equipment case is discouraged, however, since it invites trouble in the form of ground loops or audio hum noise and short-circuits phantom power supplies, which will be explained later. The cable shield, if connected, is always connected to pin number 1. When line-level (high-level) audio is fed through XLR-3s, pin number 1 or shield is connected only at the source end of the cable to minimize ground loops. And in unbalanced cables, that is, one-wire shielded cables, the cable's ground or shield (outer braid) must be connected at both ends to complete the signal path.

Regarding audio connectors and audio standards for them, practices are currently in use that also include levels, connectors, and wiring. The first standard maintains that the older, broadcast reference of zero VU is equal to +4 dB into a 600 ohm balanced circuit, typically using XLR-3 connectors. Pin number 2 of the connector is the electrical high side of the balanced circuit and is connected to a red color-coded wire. Pin number 3 is the low side of the circuit, with pin number 1 as ground, completing the balance.

The importance of this is twofold. First, if plugging a mike into a cable that is connected to equipment causes a loud whine, buzzing sound, or crackling in the system, with perhaps no audio signal, there is an open (broken, disconnected) or inadequate ground. The wire to pin number 1 in the cable may be disconnected at one of the ends, often at the pin. It is possible that the "open" is in the mike or in the receptacle, but it is more likely to be in the cable, and at the connector, because of the hard physical use all mike cables get daily. Poorly designed input amplifiers may also cause these noises. Second, if a low hum is heard when the mike is plugged in, a ground loop may be present.

A hum is normally an inductively induced sound. If the circuit is balanced, ground loops can usually be eliminated by lifting an audio ground at the source end. It is also possible that one wire of an audio pair in a balanced line is disconnected or is passing through a high DC resistance that has effectively unbalanced the line. Unbalanced lines, particularly lengthy ones, are frequent receptors of inductive noise, such as hum. Inductive noise can be avoided by routinely running audio cables several inches away from computer data lines, telephone company dial lines, and any AC power lines, especially the SCR lighting dimmer cables found in television studios. It is a good idea to avoid audio cable proximity to fluorescent lamps for the same reason.

Additionally, should a cable's XLR-3 connector have been wired with pin numbers 2 and 3 reversed in polarity (the wrong wire to the right pin), then the mike connected to that cable and connector would be electrically "out of phase" with any other mikes connected to the same console. The consequences of phasing error will be discussed in detail later.

The older, heavy-duty mike cables had three flexible stranded copper wires, each covered with rubber and usually color-coded white, black, and brown. They were surrounded by a braided copper shield that protected against the intrusion of external electrical fields. The shield was covered by a rugged black or brown neoprene or rubber outer cover. The newer mike cables have two inner wires, usually color-coded red (pin number 2) and black (pin number 1), surrounded by an aluminum foil shield, with the outer cover in various colored rubber, neoprene, or fabric material.

There is no formal standardization of color-coded wire to XLR-3 pins, except as mentioned earlier, but the important thing to remember is that within a given audio system or installation, standardization must be maintained from the outset to avoid inadvertent phase reversal.

The general rule is as follows: with a three-wire cable, connect the white wire to ground (pin number 1) and the other two wires as appropriate to the system. What is appropriate to the system may be easily checked by releasing and sliding back the shells of two other mike cables' male ends (easier to work with than the female end) enough to expose the wiring color code. Examine two connectors to avoid a chance error.

In a two-wire shielded cable color-coded red and black, connect red to pin number 2, black to pin number 3, and shield to pin number 1 and to the shell connector.

Problems occur in mike cables sustaining hard use because despite the strain relief built in at the connector's cable end, there is often too much individual wire slack within the connector shell. The wires to individual pins should never be longer than ¼inch or should only be long enough to reach to the inside of the connector pins.

We give this connector and cable information in such detail because of the consternation caused when hum, crackling, or phasing problems occur.

Although the first standard, with pin number 2 "high," is recognized by professional broadcast technicians in the United States, it is almost universally ignored by recording studios and by the semiprofessional audio industry. A safe practice is to obtain the manufacturer's literature, especially where semipro equipment and both balanced and unbalanced audio devices are interfaced.

The second standard of practice, which has its roots in semiprofessional or "hi-fi" audio systems, mostly music systems, has become an alternate standard in some studios. This is due both to the appearance in the industry of the musician-technician and to integrated circuit chip technology.

In this standard, zero VU can be anywhere from -10 to $+4$ dBv and into loads from 600 ohm up, although 10 Kohm, or what is considered bridging, seems to be the average impedance used. Bridging is discussed in Chapter 8.

Typically, semipro audio systems use a TRS (tip-ring-sleeve) connector (Figure 5–3) rather than an XLR. This is the type of connector normally found at the cable end of stereo headphones. It is ¼ inch in diameter and has a tip, a ring, and a sleeve as contact points, with the tip as the high side of the circuit, the ring as the low side, and the sleeve as the shield or ground. If XLR-3s are used in this second standard, pin number 3 will most likely be connected as the high side, as in European systems.

Where semipro equipment, called IHF equipment, must be interfaced with professional audio gear, a device called The Matchbox (Figure 5–4) may be used. The Matchbox is a 2 lb, 6¼ inch × 3¾ inch × 2¼ inch bidirectional unit with four independent amplifiers providing stereo input and output interface. Two amplifiers convert a stereo IHF, high-impedance unbalanced source to low-impedance balanced outputs at studio level. A second pair of amplifiers converts a stereo-balanced studio line source to unbalanced IHF outputs to feed the inputs of an IHF device.

RECORD TURNTABLES AND DISK REPRODUCERS

A large part of a radio station's broadcast output is music, for the most part derived from long-playing (LP) record albums, 45s, and CDs. The control operator mixes these recordings into the program by playing them on disk reproducers called turntables or on CD players. Stations with a high music-content format have two or even three turntables and at least two CD

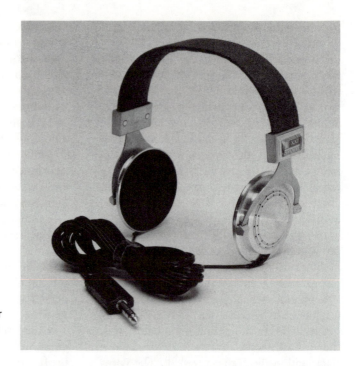

Figure 5–3 The TRS connector on Stanton earphones. (Courtesy Stanton Magnetics, Inc.)

players in each control room so that music can be played without interruption.

Modern broadcast turntables are usually designed to play 33⅓ and 45 revolutions per minute (RPM) records, with many still able to play at the now-discontinued speed of 78 RPM. Stations still carrying old 78s in their record libraries often have one table available at that speed for dubbing (copying) the 78s to tape or cart before use on the air.

The reader should visualize a turntable as being composed of three systems: the mechanical platter, the tone arm, and the phono cartridge-stylus.

The turntable platter is a heavy metal plate with a round pin projection in its center to fit the record center hole. This

Figure 5–4 The Matchbox. (Courtesy Henry Engineering.)

plate is felt or rubber covered and is mounted over a speed-change gear mechanism and a drive motor. The motor drives the plate at a fixed speed directly, by belt drive, or by a rubber-puck drive against the plate's inner rim. A turntable's quality is in part measured by the lack of "wow and flutter" in its mechanical drive that can be intercepted or picked up by its transducer and made part of the audio signal. Wow, or, to be accurate, lack of wow is the turntable's ability to come up to speed, and lack of flutter is its ability to remain at speed without the slightest variation. Wow is thought of as low-frequency and flutter as high-frequency pitch variation.

The other turntable quality factor is called rumble. Rumble is a low-frequency sound caused by one of many possible factors, such as vibration of the earth, vibration of the turntable mounting, or any other vibration that could cause the transducer to pick up low-frequency sound and mix it with the audio signal.

The control operator operates the turntable motor. The motor has an on-off switch and speed-change device mounted on its cabinet within easy reach. Alternately, the motor can be started by logic controls on the turntable's input on the console.

The tone arm is a short rod, often S shaped, that is pivoted at its mounted end with a counterweight beyond. This counterweight allows adjustment of its weight at the other end, which carries the phono cartridge. Adjusting the counterbalance weight sets the vertical tracking force (VTF) of the stylus pushing down on the record groove. Opposing this VTF is the spinning of the record and the groove modulations that are pushing the tone arm up, reducing the VTF. Tone-arm design, then, aims at a balance for constant VTF and damping at the pivot so that the arm will float rather than fall to the record surface. Additionally, tone arms are de-

signed with lateral balance adjustment for tracking warped records and for anti-skate. Skating is the tendency of a light tone arm to slide across the record, toward the center of the record disk, with the stylus destroying the groove surface by scratching across the grooves. The tone arm additionally must not be able to resonate (vibrate) in and of itself at any audio frequency that can be picked up by the transducer and become part of the audio signal.

The phono cartridge is the transducer, and the stylus (needle) that rides in the record groove is its activator. Phono cartridges are either moving coil or moving magnet types. The transducer operates as the stylus affects the motion of a magnet in a coil or the motion of a coil around a magnet. The coils or magnets in a phono cartridge are located in the ideal geometric spots to extract information from their side of the record groove while ignoring signal from the other groove wall. This design provides stereo separation and lowers the effective moving mass seen by both sides of the groove. The result is better frequency response, better tracking, and less record wear.

The stylus rides the groove wall and is caused to vibrate. It imparts the vibrations to the coils or magnets in the phono cartridge, which transform them into minute audio signal, left and right channel. A preamplifier in close proximity builds the signals to line level.

The stylus is constructed of industrial diamond (biradial, elliptical, or conical) bonded to a beryllium shaft for rigidity. It eventually begins to show wear caused by friction, and it is easily replaced by the audio operator before it develops flats, which will injure the record groove and audio quality.

A discrete equalizer, which in the past was built into the turntable, with control switches available to the operator, is now found controlless, built into the turntable preamp. This preamp is either in the turntable or in the console's turntable-input channel. The equalizer is used to compensate for worn and distorted record groove and to match the equalization curves of record manufacturers (Record Industry Association of America [RIAA] and National Association of Broadcasters [NAB] curves for United States manufacturers and CCIR, IEC, and DIN curves for European manufacturers). The curves are frequency-control curves, designed among other things to limit the excursion of the cutter stylus in the recording process. They must be matched on playback to obtain the full frequency range of the music recorded.

QRK 12C Turntable

The QRK model 12C turntable (Figure 5–5) is made by Broadcast Electronics. It has

three speeds, 33⅓, 45, and 78 RPM, and is rim driven; speed change is accomplished by a gear-shift lever on the front. Its start-up time for all three speeds is ⅛th revolution for full speed. Its rumble is rated at −38 dB against the NAB standard of −35 dB. Wow and flutter are less than 0.1%.

EMT Turntables

EMT turntables are made in Germany. The EMT 938 (Figure 5–6) is a three-speed turntable with the following controls, from left to right: power on-off, start-stop, tone-arm lift, and speed switch. A stereo phone jack is on the extreme right. Run-up of the turntable platter and platter braking are accomplished in less than 500 msec. A professional turntable is expected to provide fast fade-ins, so a short run-up time is an important require-

Figure 5–5   The Broadcast Electronics QRK 12C turntable. (Courtesy Broadcast Electronics, Inc.)

Figure 5–6 The EMT 938 turntable. (Courtesy Gotham Audio Corporation.)

ment. During run-up, however, considerable counterforces arise between the turntable platter and its chassis, leading to rotational oscillations around the axis of the platter. Although a dynamically balanced tone arm can compensate for purely lateral or vertical movement, it cannot do so for rotational movement. Rotational oscillations can cause tracking disturbance and wow and flutter effects, particularly during start-up. The EMT 938 is constructed to dampen such oscillation. The direct-drive system is the same as in the EMT 948. The turntable is delivered with the EMT 929 tone arm and an empty cartridge shell. It has self-contained control electronics and amplifiers. The amplifiers include standard filters and EQ and terminate in XLR-3 connectors.

The EMT 948 (Figure 5–7) features the EMT 929 tone arm and direct drive with fast start, fast stop, and reverse. It has a motor-driven tone-arm lift and built-in amplifiers with line-level outputs and an output for the cue circuit. The 948 has a plastic dust cover. It operates at all three speeds with a quartz-controlled accuracy

of $\pm 0.1\%$ and reaches full speed at a maximum of 0.5 second. Its wow and flutter at 33⅓ RPM is a maximum of $\pm 0.075\%$, and its rumble signal-to-noise ratio is 50 dB. The controls, left to right, are reverse speed, start-stop, tone-arm lift, and speed-control switch. Reverse speed is for hands-off backtrack cuing.

The EMT 950 (Figure 5–8) is a lightweight turntable that is braked and released electromagnetically from a heavy high-mass flywheel below. The turntable has a built-in large-hole adaptor for 45 RPM records that, when engaged, automatically changes the speed to 45 RPM. This model is a three-speed, electronically controlled, direct-drive system.

All the operating controls are push buttons, designed to be operated with the left hand while the right hand remains free from the manipulation of the tone arm and record. The push buttons are arranged in use sequence. From right to left are the power on-off switch, with 33⅓ to the right and 45 and 78 above, with stereo and mono switching next above. Centrally located is a perforated panel with

Figure 5–7 The EMT 948 turntable. (Courtesy Gotham Audio Corporation.)

cue loudspeaker for the turntable. To the right of the cue speaker are its level control, the motorized reverse control of the turntable for cuing, and the phono cartridge illumination control. Below these are the turntable local-remote selector, a pilot indicator for when the console fader is on, turntable start-stop controls, and the motorized tone-arm lift-lower control.

Wow and flutter are a maximum of ±0.05%, and the rumble signal-to-noise ratio is 56 dB.

Tone Arms and Phono Cartridges

Tone arms and phono cartridges are often purchased separately from the turntables on which they are used.

The Audio-Technica model AT 1010 tone arm in Figure 5–9A is designed so that the pivot point is at or below the plane of the record, thus counteracting the tracking forces generated by the record's motion. The ATP 12T (Figure 5–9B) and ATP 16T (Figure 5–9C) are stereo tone arms.

The Audio-Technica AT30E (Figure 5–10) is a stereo phono cartridge with moving coils. Controversy exists in the industry over whether moving coil or moving magnet phono cartridges are superior. Most manufacturers make both types, so perhaps the advertising factor is part of the picture.

The AT30E has a frequency response of 15 Hz to 25 KHz, channel separation of 25 dB at 1 KHz, biradial stylus shape,

Figure 5–8   The EMT 950 turntable. (Courtesy Gotham Audio Corporation.)

a tracking force of 1.4 to 2 g, and a cartridge weight of 5 g.

The Stanton 500 broadcast series cartridges (Figure 5–11) have a frequency response of 10 Hz to 22 KHz, a 35 dB channel separation, an elliptical diamond stylus, and a tracking force of from 0.75 to 1.5 g. The cartridge weighs 5.5 g and is rated by its manufacturer as having high compliance, low mass, and superb tracking ability.

The Shure V 15 type V (Figure 5–12) has a response of 20 Hz to 20 KHz, channel separation of 25 dB or greater at 1 KHz, hyperelliptical stylus shape, and a tracking force of 1 to 1.25 g, plus 0.5 g for the dynamic stabilizer, which is a built-in brush that cleans and destaticizes

the groove surface and protects the stylus from shock. The cartridge weighs 6.6 g.

COMPACT DISK PLAYERS

Playback of CDs begins with loading a CD into the player. Two loading systems are used in CD players. First is the type where the user places the disk into an open drawer and the drawer is automatically pulled into the machine. The disk is then lowered onto a drive spindle and is clamped tight at the disk center in a process known as chucking. It is similar to the way that a drill bit is clamped into a drill chuck.

The second type of disk loading, top

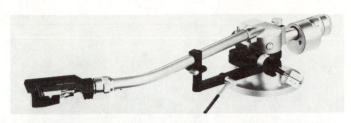

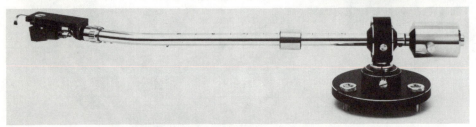

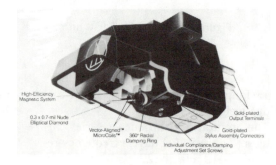

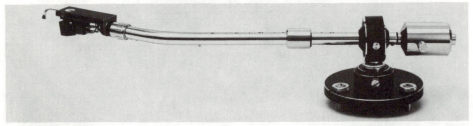

Figure 5–9 The Audio-Technica AT1010 *(top)*, ATP-12T *(middle)*, and ATP-16T *(bottom)* tone arms. (Courtesy Audio-Technica.)

Figure 5–10 The Audio-Technica AT30E phono cartridge. (Courtesy Audio-Technica.)

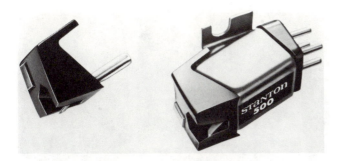

Figure 5–11   The Stanton 500 phono cartridge and stylus. (Courtesy Stanton Magnetics, Inc.)

loading, has the user place the disk directly on the drive spindle, and the chuck or clamp is attached to a lid that must be closed before the device can play. Both loading systems use a safety switch, which prevents the laser beam pick-up from operating and emitting light if the lid or drawer is open. These precautionary measures ensure that the user's eyes will not be exposed to direct laser light.

In the player, the data path includes the pick-up laser, the data separator, the time-base corrector, and the error-correction process and error-concealment device. The sample data that result are fed to the convertors, which supply the music information back to the analog world.

The microprocessor control system in use has operator controls consisting of play and eject, a pause control, and most

Figure 5–12   The Shure V15 type V phono cartridge. (Courtesy Shure Brothers, Inc.)

likely controls that permit moving the pick-up laser very rapidly from track to track on the disk as well. CD players designed for broadcast use have a control for cuing. On some of these machines, a hand-operated rotor simulates the turning of a record manually and locates the cue-in point. After cue-up, the user operates the start control, starts the player, and brings in modulation with the CD player input pot on the console.

Denon Model DN 950F

The Denon model DN 950F CD cart player (Figure 5–13) was designed to blend in with the station's cart machines. The CDs are preplaced in plastic cartridges, as shown below the machine in Figure 5–13. The cartridge protects the CD from scratches and dust particles, and the CDs in carts are loaded directly into the player. A shutter on the cartridge opens and closes on loading and unloading. Below the loading slot on the DN 950F is a digital display that reads out the CD track number and remaining time in minutes, seconds, and frames. An end-of-message (EOM) signal is given before completion of a selection. The digital display blinks when a CD is loaded improperly or when disk reading is incomplete.

Track selection on the CD is performed with the rotary pulse encoder dial, at the lower left of the panel. The user turns the dial in either direction the desired number

Figure 5–13   Denon 950F CD cart player. (Courtesy Denon America, Inc.)

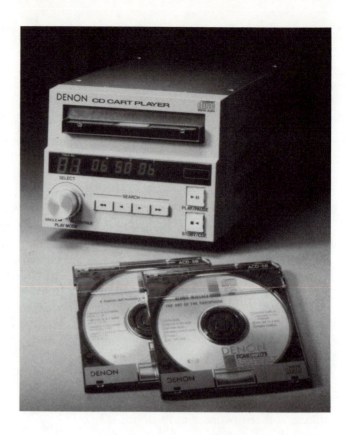

of clicks to select the assigned track number. The pick-up follows instantly while the display indicates the exact pick-up position. One full rotation covers 20 tracks on a CD. Turning clockwise increases track count and counter-clockwise decreases track count. When the dial is turned counterclockwise past the first track, the last track number on the CD appears on the display.

Concentrically below the dial is the play mode switch, which can be set to single or continuous play modes. In single play mode, the playback stops after completing a specific single track. In the continuous mode, subsequent tracks are continuously played, and the machine stops on completing the final track on the disk. The search buttons in the lower center are used for forward and reverse searches. Pressing one of the inner buttons advances the optical pick-up in increments of one frame. Pressing one of the outer buttons advances it in increments of about 0.5 second. Holding a button down will allow the pick-up to scan continuously and faster.

On the lower right of the panel, the play-pause button, when pressed once, starts the machine and illuminates the red indicator light. When the button is pressed a second time, the machine pauses and the yellow indicator light comes on.

The standby-cue button, below play-pause, when pressed during play, recues the pick-up instantly to the last point where the play button was pressed and engages the stand-by mode. If the button is pressed during repeat play, when the search function is engaged, the pick-up will enter the stand-by mode at the present selection.

Sony CDP 3000 CD Player and CDS 3000 Control Unit

The Sony CDS controller (Figure 5–14) remotely controls two CDP players, which provide continuous play time. The player selectors at the upper right of the controller each have illuminated indicators showing the number and operating mode of the corresponding player. Two players may be mounted side by side in a standard 19 inch rack. The only control on the player is its eject button. The CD disk is placed in a drawer on the front of the player. The programming instructions that follow are performed as the user looks at the face of the CDS 3000 controller.

1. To preset selections on the controller, press the program key. The program 1 indicator starts flashing. Select the player and put it in the manual or monitor play mode. Find the desired music selection using the rotary search dial or the ten keys to its right. Press the set key. The program 1 indicator stays lit, and indicator number 2 starts flashing. The set key is operative only when the selected player is in the manual or standby mode. Repeating all the above steps provides up to eight selections from two disks, which can be randomly preset. For fewer than eight selections, press the program key after the last selection.

2. To run the program of preset selections automatically, press the auto-manual key. The start point of the first selection will be automatically searched, and the standby indicator will light. To play, press the play key for on-line play or the monitor play key for monitoring, to start the first selection. The program 1

Figure 5–14   Sony CDP-CDS 3000 CD Player. (Courtesy Sony Corporation.)

indicator flashes. When the first selection ends, the player goes to stand-by mode at the beginning of the next preset selection. Press the play or monitor key to play each selection in turn.

The upper left of the controller has a time display that, in the manual or stand-by mode, shows the location of the pick-up's track number, index, and time in minutes, seconds, and frames. During play, when the remain time key is pressed, the display is switched to indicate the remaining time of the music track being played.

TAPE PLAYBACK

Tape playback is the second most important tool in the audio operator's hand, with only console itself taking first place. Tape's flexibility of use in delayed broadcast of special events, its ease of editing, its use in structuring the recording of music, and its ability of erasure and reuse all make it an ideal record-reproduce medium. Of course, one cannot discuss tape play-back devices without acknowledging that they are also recording devices, and so their recording function will also be discussed.

The three distinct types of analog tape play-back equipment in general studio use are open reel, cartridge, and cassette. The latter, analog cassette although not originally designed as broadcast equipment, has because of its small size and simplicity of operation been refined to increase its level of quality and has insinuated itself into broadcast operations.

Open-Reel Tape Playback

A professional analog open-reel tape machine has two integral sections or parts: the tape transport, or puller, and the electronics, or logic, section. For playback the

electronics section of the tape machine feeds the output of the playback head(s) to the control console's tape input(s). The puller moves the tape from a feed reel on the left across the tape heads to a take-up reel on the right at a constant speed. The constant speed is maintained by one of two systems. The first and most commonly used is the capstan and pinch roller system—a pair of tangent rollers, the capstan, metal, and the rubber or neoprene pinch roller—through which the tape must pass as it moves across the heads from feed reel to take-up reel. The second system for maintaining constant speed employs a servomotor mechanism and a sensor that operates an oscillator. The sensor senses minute speed changes and "tells" the oscillator, which speeds up or slows down the motor, controlling the speed of the tape across the heads.

The playback has a speed and equalization control that can put the machine in the 3.75, 7.5, or 15 inches per second (ips) speed, and it has a mode switch that can be positioned to play, record, fast forward, fast rewind, or stop. The electronics section of the machine contains a playback amplifier, a record amplifier, an eraser oscillator, a bias oscillator, the logic functions as well as the VU meter. Equalization is provided for each of the machine's speeds, and monitor functions are provided for input and output modes.

The tape heads are arranged in a line on the puller between the two reels. In a basic tape machine, they are arranged in the following order: erase head, record head, playback head. The azimuth settings of the record and playback heads are critical and must be readjusted from time to time. In multitrack machines, the heads are in stacks of erase, record, and playback.

Open-reel analog tape machines for broadcast and recording studio use operate at any one of four standard speeds: 3.75, 7.5, 15, or 30 ips. The latter two speeds are used mainly by the recording industry for maximum ease in editing and maximum quality of reproduction. Current open-reel machines often provide for slight variation of standard speeds to permit pitch change adjustment. Tape speeds of $^{15}/_{16}$ and $1^{7}/_{8}$ ips are used in cassette recorders.

The quality of recording attributable to the recorder, which affects the quality of reproduction on tape, is a function of tape speed, track width, and head-gap width. Quality, which is measurable as frequency response and signal-to-noise ratio, increases as tape speed is increased, the head gap (the space between the pole pieces of the head electromagnet) is narrowed, or the tape track is widened. Head-gap width can be narrowed only so much before the pole pieces touch.

The track is the magnetic signal that is layed down by the record head on the tape, in the direction of travel of the tape. Figure 5–15 shows a tape path. A monaural recording has a track that is virtually the full width of the tape. A multitrack recording uses a small portion of the tape width, with a space between tracks, for each recorded track. The width of each track depends on how many tracks are recorded on a given tape width.

As the number of tracks is increased, there is 3 dB of added noise per track. This noise is inherent in the electronic properties of both the tape and its amplifiers. The noise level increases because, with a greater number of tracks on a given tape width, each track becomes narrower, requiring greater amplification; as the amplification increases, the inherent noise increases. Additionally, as the track narrows, the granulation size of the magnetic oxide formulation begins to play a part. The process may be likened to enlargement of a photograph from a small negative. As the enlargement gets progressively larger, the grain (the noise) in the photographic print becomes larger and more apparent.

Recorded level on tape is measured by its magnetic properties and is stated in nano-Webers per meter, or nW/m.

To operate any analog open-reel tape machine, place a reel of tape on the feed spindle and secure it. Pull the end of the tape through its tape path around the tape tension idlers, across the heads, and between the capstan and pinch roller, and attach the tape to the take-up reel. Switch the machine to the desired speed. To play a recorded tape, switch the machine to the play mode, start the machine, and open the machine's output pot if there is one. To record, first feed modulation to the machine, open the machine's input pot(s), and adjust the level. Switch the machine to the record mode, start the machine, and control the input gain.

Figure 5–15   Tape path.

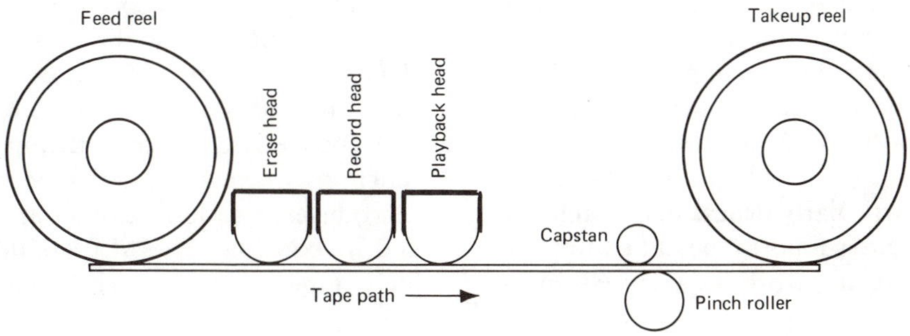

Digital Audio Tape Recorder

The two distinct types of digital audio tape recorders are those with fixed, or stationary, record and playback heads, called DASH (digital audio stationary head) machines, and those with rotary record and playback heads, called R-DAT (rotary head digital audio tape) machines.

At first it may seem that the rotary head machines have all the advantages, but one must look closer. Rotary head machines are used primarily for mastering tapes for CDs and in consumer-level R-DAT or just DAT tape machines. The rotary head machine can provide the high head-to-tape speed necessary for high-density (2 megabits per second) recording of a stereo signal, combined with reasonable tape consumption.

High head-to-tape speed is desirable and can be accomplished in one of two ways in a tape machine. If the head is held stationary and the tape is transported very rapidly, high head-to-tape speed is accomplished with a very large amount of tape expended. If both the tape and the head are moving at moderate speeds, then the high head-to-tape speed is accomplished using much less tape. This second method is the one used in rotary head recorders.

Rotary head recorders may use either the transverse scan or helical scan method of reading information from, and writing information to, the tape.

Although rotary head recorders can achieve a higher information density and use less tape than the stationary head machines, the rotary machine has bandwidth problems in multitrack applications. A 36-track recorder, for instance, would need a bandwidth of 36 megabits of information per second. Further, rotary head machines, particularly helical scan machines, are unable to support tape-cut editing. Of course, in audio work, tape-cut editing is the type most often used because editors have used it since audio tape's beginnings.

Therefore, multitrack digital audio recorders, those used in the primary recording of music, have stationary head stacks and open-tape reels, and they very much resemble analog recorders.

Digital-Analog Interfacing

Since most of the audio world is still, at this writing, in the analog format, we must be able to interface the outputs of digital devices to analog systems and, indeed, to other digital devices. Particularly prevalent is the interface of digital recorders to analog consoles.

At the outset of digital, each manufacturer devised its own, and incompatible with others, interfacing system. This proved intolerable, and standardization was reached using what is now known as the AES/EBU standard. (AES is the Audio Engineering Society, and EBU is the European Broadcasting Union.) The standard embraces all the functions used by previous interconnect formats and is independent of equipment manufacturer.

Some examples of analog tape recorder-reproducer open-reel models in use in the industry are discussed below.

Various Open-Reel Models

Otari open-reel tape machines are manufactured by Otari Electric Co. LTD of Japan. We will look at the MX 5050 B-II (Figure 5–16), the 5050 Mark 111, and the MTR series of master tape recorders.

A good way to describe the operational functions of an open-reel tape recorder-reproducer is to look closely at its puller and its controls. The MX 5050B is a ¼ inch tape, two-channel, quarter-track

Figure 5–16   The Otari MX-5050B-II tape machine. (Courtesy Otari Corporation.)

(with full track optional) recorder-re-producer that has an additional quarter-track switch-selectable stereo reproduce head. Selection between the two heads is made by a switch accessible through the top of the head cover. The machine uses reels of up to 10½ inches.

Between its two reels is a pitch and off-speed control to correct prerecorded tapes that are slightly off speed or to synchronize the pitch of previously recorded music with the pitch of instruments to be recorded on the other track. It provides a speed change of ±7% from nominal speed if the user pulls its knob up and turns it clockwise or counterclockwise to increase or decrease speed and therefore pitch. A red LED indicator next to the knob illuminates when the control is operative. When the knob is pushed in, the LED is extinguished and the machine is returned to its nominal speed.

Below the feed reel an adjustable cue control helps locate tape selections by de-feating the head lifters in fast forward or rewind modes, allowing the tape to contact the head assembly in these modes and enabling a high-pitched squeal signal. The head lifters normally lift the tape away from the heads in these modes to prevent excess tape and head wear. The cue control also lowers the signal level to prevent overload of the earphones or monitor speakers.

Below the cue control is the selection locator memory and tape counter. In operation, the memory works with the counter to stop the puller when the counter passes 0000, thus allowing fast recuing to whatever point on the tape has been set at 0000. To the right of the counter is the tape head assembly. Its head cover has a tape-splicing block built into its face.

Below and to the left of the heads are a set of four push switches that select power, on-off function; speed, high or low (the MX 5050B has three speeds available in either of two pairs—15/7.5 or 7.5/3.75 ips); reel size, large or small; and edit.

When edit is selected, a green LED to the right of the edit button lights. In edit mode the take-up motor is deactivated and the end-of-the-tape tension arm switch is inoperative, allowing the tape to spill freely between the capstan and the take-up reel. After the edit point is found, the edit is made on the splicing block, on the head cover.

Below these selector switches are the two input-level controls, concentric, mike and line, rotary-level controls that may be mixed. To their right are the two VU meters, one for each channel. Each meter has a red LED peak indicator as an integral part. The peaker is normally set to flash at +9 dB, but this threshold can be adjusted internally (i.e., within the recorder).

To the right of the VU meters is a concentric, rotary, dual-output-level control. This dual control can be defeated by a standard reference level (SRL) switch located directly beneath it. When the output controls are lifted from the circuit, a +4 dBm signal is fed to any balanced 600 ohm line. Additionally, the +4 dBm can be changed to a −10 dBm to accommodate interfacing with lower-level equipment inputs. The level change is accomplished by a switch on the bottom panel of the machine.

The mode switches, above the right VU meter, are record, play, stop, rewind, and fast forward. To record, depress the play button before the record button. Additionally, depress either or both of the channel record buttons (CH 1/CH 2) located at the extreme lower left below the input controls. When either of these red buttons is depressed, a red LED above the switch is illuminated.

An interesting feature of the MX 5050 B-II is the selective reproduce (Sel Rep) switches for channels one and two located

to the right of the channel record switches. These Sel Rep switches permit synchronous playback through the record head of the channel not recording, allowing the user to record on the other channel in time synchronization with the modulation monitored on the Sel Rep channel.

To the right of the Sel Rep switches and across the base of the panel are the front-adjustable bias and record-equalization screwdriver-operated controls. These are followed by an earphones jack, the test oscillator momentary push buttons for 1 KHz and 10 KHz tones, and two monitor select push buttons, one for each channel and each able to select source or tape input.

Motion-sensing control logic permits the MX 5050 B-II to switch directly into play from fast forward or rewind without time delay, breaking or stretching the tape, or having to first go to stop. The logic system also provides freedom from recording clicks on the tape when the user punches in or out of record while the machine is running.

The bottom panel of the machine, in addition to the previously mentioned output-level adjust control, has XLR-3 connectors for mike and line inputs to each channel, line outputs from each channel, a remote control connector for remote operation of the machine's modes, a mike attenuator 20 dB pad, and a switch providing a choice of three record levels. To maximize performance with different tapes there are three separate calibrated record-reference levels: 185 nW/m (original tape standard), 250 nW/m (for high-output, low-noise tapes), and 320 nW/m (the IEC and DIN standard). Rewind time is less than 90 seconds for a 2400 foot reel.

Model 5050 BQ-II (Figure 5–17) is essentially the same as the just-described MX 5050 B-II, except that it is a four-channel recorder with separate mike- and line-input controls for each channel, in-

Figure 5–17 The Otari 5050BQ-Series II. (Courtesy Otari Corporation.)

dividual VU meters, and headphone monitoring for each channel with selectable combinations.

The Otari 5050 Mark III/4 (Figure 5–18) is a four-track (four-channel) re-

corder-reproducer that uses ½ inch tape. Tape handling, including dynamic braking, is handled by microprocessors, which also operate transport logic and motion sensing. The capstan is servomotor con-

Figure 5–18 The Otari 5050 Mark III/4. (Courtesy Otari Corporation.)

trolled, and this machine is switchable to operate at either 15 or 7.5 ips. The Mark III/4 can be interfaced to tape machine controllers and synchronizers and is compatible with dbx noise reduction systems. Optional remote controllers that duplicate the operational controls are available.

The Mark III/4 has a ±7% variable speed control, selective reproduce for overdubbing, and built-in dual-frequency test oscillator. The transport modes are operated on the puller deck, while the input and monitoring controls are located on an overbridge above the puller.

The Mark III/8 (Figure 5–19) is an eight-track recorder-reproducer using 0.5 inch tape. It has all the features of the Mark III/4, with its monitor functions located on the machine's vertical front and duplicated in a remote controller, the CB-110, which handles transport controls, channel switching, varispeed, and remote

electronic timing. Shown also, mounted above the CB-110, is the CB-116 autolocator, which offers six cue-location memories and a record punch-in foot switch. The Mark III/8 operates at 15 or 7.5 ips. It has an interface connector for Society of Motion Picture and Television Engineers/European Broadcast Union (SMPTE/EBU) time-code synchronizers and editors.

The Otari MTR 10 recorder-reproducers are microprocessor-controlled ¼ and ½ inch tape machines available in ¼ inch two-channel, ½ inch two-channel, ½ inch four-channel, and an optional ¼ inch two-channel with SMPTE/EBU center track formats. The four-channel model, which normally is a ½ inch machine, can be changed to ¼ inch tape with a conversion kit. All versions are available unmounted for custom installation or may be ordered in either of two rollaround desk consoles. The two-channel model in Figure 5–20

Figure 5–19   The Otari MX5050 Mark III/8. (Courtesy Otari Corporation.)

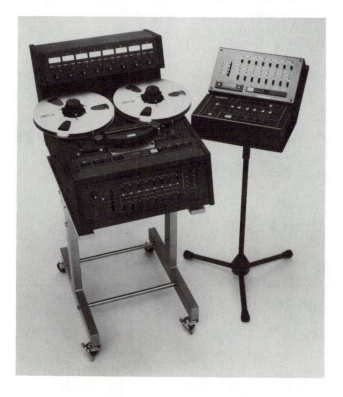

Figure 5–20 The Otari MTR 10/2 two-channel recorder-reproducer. (Courtesy Otari Corporation.)

and the four-channel model in Figure 5–21 are mounted in the low-profile cabinet. A second mounting version has the meter panel above the puller in an over-bridge. Both models feature a hinged transport assembly that can be tilted back for ease of maintenance access.

The input-output circuit boards are

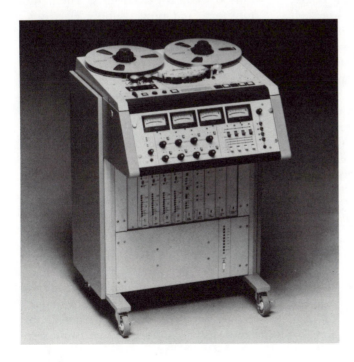

Figure 5–21 The Otari MTR 10/4 four-channel recorder-reproducer. (Courtesy Otari Corporation.)

modular, plugging into a card frame below the puller. Transport-control electronic circuit boards are located in the card frame as well, and the modular power supply is located below.

An automatic-reel size sensor determines the puller tape tension depending on reel size. Below the feed reel are two large LED readout displays. The upper display reads tape speed in ips or as percentage of nominal speed, depending on selector push button. A push button switches the display on or off.

The second display, below, is a tape timer showing real time for the three speeds of 30, 15, and 7.5 ips. The speed-select switch with automatic equalization is located on the right of the puller, below the take-up reel. The speed-mode switch is located directly below the displays. This switch chooses fixed speed, variable speed, or external control. To its left are the forward and reverse edit control and the cue control, which defeats the tape lifters. To its right are the edit-unload button, the splicing block, and the transport mode switches. Threading the MTR 10 is fast and straightforward. Once it is threaded, a lighted stop button on the puller will flash for confirmation. Touching this button will then arm the servo system, and tape is ready to roll. All the transport controls are dual lighted, with the brightest control confirming transport mode.

The audio control panel is mounted either on the overbridge or directly below the puller, depending on the enclosure chosen. It contains the VU meters, the record-level reference calibration (switchable to 185, 250, or 320 nW/m, depending on the tape used), the test oscillator control, and the input and output controls, each with its LED indicator and switch for standard reference level.

To the right are each channel's safe-ready-record switches, with LED indicators, and below are the switches for each channel's monitoring of input, Sel-Rep, or

reproduction (playback), together with their LED indicators. Last are the earphone level control and channel combination switches.

Three remote control units are optionally available for the MTR 10 machines to allow access to transport, variable speed, and autolocate functions.

Otari's model MTR 90 II is an 8-, 16-, and 24-track record-reproducer. Microprocessor controlled, it is a pinch-rollerless master multitrack recorder. It is available in both 1 inch and 2 inch tape-width transport configurations. The MTR 90 II/8 is the 1 inch, eight-channel version. The MTR 90 II/16 is the 2 inch, 16-channel version, and the MTR 90 II/16-24 is the 16-channel version, which is upgradable to 24 channels. There is also the MTR 90-II/24, which is the 24-channel machine, convertible down to 16-channel operation, and this version is shown in Figure 5–22.

All audio electronics are contained on single-channel, single-card, plug-in circuit boards. A separate card frame holds transport control electronics. The audio, transport, and power-supply electronics are cabinet mounted below the VU meters and are accessible from the front of the machine through hinged doors.

The MTR 90-II operates at 30 or 15 ips. The speed-change switch is on the left side of the head housing, with the digital tape counter in the middle and its reset switch on the right. A built-in variable speed oscillator offers a ±20% speed variation. Should an external speed control source be desired, the three-position speed-mode selector at the lower left of the top plate can select fixed, variable, or external speeds. To the right of the speed mode selector are the digital readout and pitch control knob. Beyond the pitch control is the cue button tape lifter defeat and a continuously variable forward and reverse slide control. When engaged, this slide feature allows continuous and step-

Figure 5–22   The Otari MTR-90-II multichannel recorder-reproducer. (Courtesy Otari Corporation.)

less speed control from 0 to 45 ips forward and reverse. Both precise cuing and stable slow-speed tape crawls can be achieved. Edits can be performed precisely and efficiently while the user maintains "play condition" control normally lost during manual reel cuing. Manual reel cuing can be done by turning the capstan with a finger.

To the right of the speed control slide are the dual-lighted transport-control switches. The record switch has a physical guard protecting it from inadvertent operation.

Because of the complexity of 24-channel recording, all versions of the MTR 90-II include the CB-113 remote session controller, which provides separate record, ready, and monitor functions, varispeed control, master switching, cue, and full transport controls. An optional autolocator, the CB-115, has ten assignable memories.

Scully recorder-reproducers are made by the L.J. Scully Manufacturing Corp., an old-line maker of disk recording equipment.

The LJ-12 (Figure 5–23) is an analog recorder, with digital control of all transport and audio functions. It is available in ¼ inch tape in monaural, two-track or one-quarter track and in ½ inch tape in two or four track at tape speeds of 3.75, 7.5, 15, and 30 ips. It has a variable speed range of 3 to 36 ips, in 0.01 ips increments, with digital display of either ips or percentage of deviation.

Figure 5–24 shows the bottom of the puller, which indicates that it can be tilted up for easy maintenance.

The LJ-12 may be interfaced to synchronizers. It has separate audio memories for every audio parameter and all tape speeds, and it has a built-in monitor amplifier with headphone jack and speaker terminals.

The TASCAM ATR 80-24 (Figure 5–25) is a 24-track, 24-channel recorder-re-

Figure 5–23 The Scully LJ-12, top view. (Courtesy L.J. Scully Manufacturing Corporation.)

Figure 5–24 The Scully LJ-12, bottom view. (Courtesy L.J. Scully Manufacturing Corporation.)

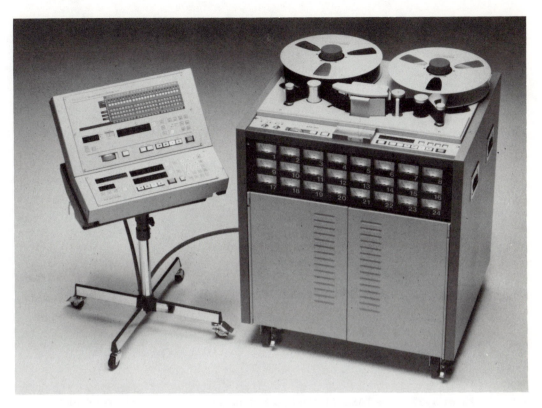

Figure 5–25   The TASCAM ATR 80-24 with autocontrol and autolocator. (Reprinted with permission from TEAC Corporation of America.)

producer using 2 inch recording tape on up to 14 inch reels. It operates at either 30 or 15 ips.

The transport controls are located directly in front of the puller and include a rotary shuttle control that moves tape back and forth at any speed from 0 to 150 ips in an easy fingertip operation and allows tight lineup of edit or cue points. Return to zero and search-to-cue locate functions are provided in conjunction with the hour-minute-second digital readout to the right of the tape-cut block. The head stack (Figure 5–26) includes erase heads, record-synchronization heads, and reproduce heads.

The head amplifiers are on individual plug-in circuit cards with access from the front of the machine. Above the amplifier cabinet are the 24-channel VU meters.

Figure 5–27 shows a front view of the TASCAM ATR 80-24.

The movable remote control unit provides duplicated transport function controls, including the tape shuttle control and return-to-zero and search-to-cue locate functions. Input, synchronization, and reproduce modes are independently switchable for all 24 channels, or all channels can be simultaneously switched to any mode by pressing one key. Independent channel settings that have been input by the operator are retained in memory and can be recalled. A sync-lock function locks any assigned channel or groups of channels to the synchronization mode for use with SMPTE time code or other sync signal.

The AQ 80 autolocator (Figure 5–28) offers ten-key entry of locate points. Up

Figure 5–26 The TASCAM ATR 80-24 head stack. (Reprinted with permission from TEAC Corporation of America.)

Figure 5–27 The TASCAM ATR 80-24, front view. (Reprinted with permission from TEAC Corporation of America.)

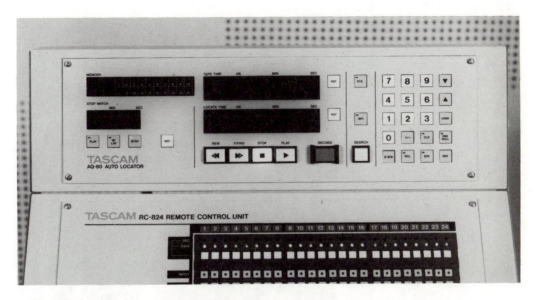

Figure 5–28   The TASCAM RC-824 remote control and AQ-80 Autolocator. (Reprinted with permission from TEAC Corporation of America.)

to 20 locate points can be stored in memory. In addition to search function, it can set up endless repeat loops as well as up to 20 seconds of preroll.

We will look at only one digital recorder-reproducer, the Sony PCM 3324, because at this writing it is the one recorder used most at major production studios and remote operation vans.

The Sony PCM 3324 (Figure 5–29) is a DASH format digital recorder-reproducer. The DASH format specifies that the recorder has stationary heads and describes its sampling frequency, tape format, linear packing density, and error correction for 2-channel to 48-channel recorders using either ¼ or ½ inch tapes.

The 3324 uses a pinch-rollerless capstan direct-drive transport under microprocessor control. Immediately below the front of the transport are 24 vertical-segment VU meters, and below the meters is a horizontal shelf that holds the machine's controls. Figure 5–30 depicts a top view of the Sony PCM 3324.

Two sampling frequencies are provided, and five points on a tape can be memorized as cue points. These cue points can then be modified as necessary and located automatically with great precision. Use of the RM 3310 remote control unit permits autolocation of up to 100 cue points as well as comprehensive automatic operations including return, repeat, roll back, and roll back to play. There is electronic crossfade, variable in 16 steps, from 1.33 msec to 372 msec, which can be timeset individually for each channel. Because the cue tracks are time aligned with digital audio, tapes that are made on the PCM 3324 can be cut edited in the same way as analog tapes. There are 24 digital audio inputs and 24 digital audio outputs, as well as inputs and outputs for synchronization, remote control, time code, and external speed control.

When the 24 audio tracks available on a single PCM 3324 are not enough, then two or even three machines can be operated in perfect synchronization to accom-

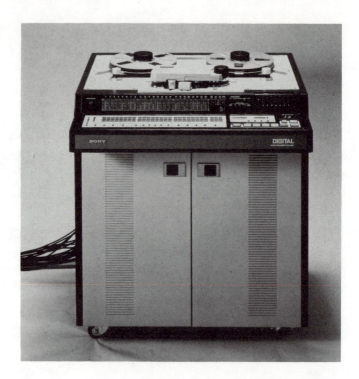

Figure 5–29   The Sony PCM 3324. (Courtesy Sony Corporation.)

modate 48 or even 72 channels. There are, in fact, 28 tracks. The other four tracks are two analog tracks, a control track, and a time-code track.

The PCM 3324 has switchable sampling rates of 48 KHz and 44.1 KHz, 16-bit quantization, and a frequency response of 20 Hz to 20 KHz, +0.5 and −1.0 dB. It weighs more than 435 lb.

Figure 5–31 shows three 3324s in a 72-channel system drawing as they would be used in a recording studio. The IF 3310 units are interfaces, making it possible for the recorder to coexist in a system with a synchronizer.

To repeat, once the audio sound or signal has been converted to digital, it is simply a series of numbers that can be

Figure 5–30   The Sony PCM 3324, top view. (Courtesy Sony Corporation.)

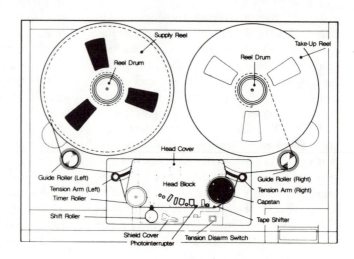

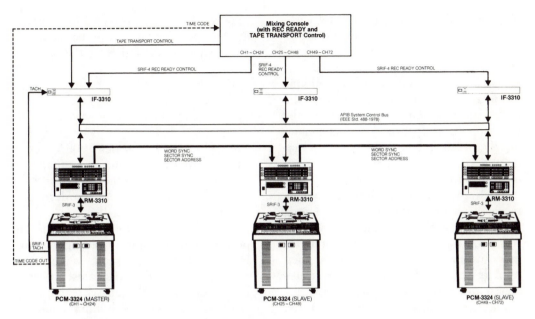

Figure 5–31   The Sony PCM 3324 72-channel system. (Courtesy Sony Corporation.)

transmitted, processed, and stored. The computer industry has perfected many ways of transmitting, processing, and storing high-speed streams of numbers, and with ever-decreasing overall cost, as the price of complex chips, memory chips, and storage media decline.

Digital audio recording can be done using either magnetic or optical computer disk technology. This technology allows fast random access to the stored information, and in audio editing, when the user is trying to locate a cue point, magnetic disk random access is much faster than spooling magnetic tape from feed reel to take-up reel and return.

Digital editing without cutting tape is done by fading or dissolving between two sources of audio samples. The edit can be previewed, as in video editing, so that the editor can hear the result and perhaps critically move the editing-in and editing-out points before the actual edit. The edit then can be recorded on a separate disk,

so that unlike cut editing, no tape cutting is necessary.

Tape cut editing, however, is supported by digital audio if stationary-head digital audio recording machines are used.

Cartridge Tape Recorders and Reproducers

Cartridge tape machines, like open-reel machines, are both inputs to and outputs from the console, depending on whether the machine is used for playback or for recording. They are usually called cart machines, and they use a continuous unending loop of specially prepared and lubricated tape enclosed in a plastic container. Carts and cart tape are discussed in Chapter 7. Cart playback has replaced record turntables for playback of recorded commercial announcements, program themes, and openings, bridges, and fanfares. Cart commercial spots have the advantage of longer reusability without measurable quality deterioration, provided that the tape heads are cleaned and demagnetized often and routinely, depending on use. Cart tape has a unique advantage over turntables—instant cue-up. A cue tone burst is recorded on the cart tape on a separate cue track immediately, precisely before the program material starts. After a cart spot has been played, the tape continues to roll silently through the cart until the cue tone is reached, which instantly stops the tape motion. The cart is then tightly cued and ready for the next play even if it is removed from the machine. It remains cued until its next use. Cart tape may be bulk-erased and reused like open reel tape.

Cart tape machines are manufactured as single-deck record-playback, or playback only, and in mono or stereo for NAB type A- and AA-sized carts. Cart sizes are discussed in Chapter 7. Cart machines are manufactured in multideck configurations of three-deck, five-deck, and ten-deck and in 40-cart carousels. The multideck machines are generally playback devices only, but often they can be ordered with one deck configured as both a record and play machine. All cart machines operate at 7.5 ips speed but can be internally strapped or jumpered for either 3.75 or 15 ips.

Broadcast Electronics Cart Machines

We will look at the series 3000A and series 5300C cart machines manufactured by Broadcast Electronics. Figure 5–32 shows the series 3000A recorder-reproducer.

To play back on the 3000A, turn the power switch on the lower right to on. Insert a prerecorded cart in the deck opening and firmly up against its right-hand side. When the cart is seated properly, the stop switch will illuminate to indicate that the machine is in the "ready" mode. Momentarily depress the start switch. The tape will run, the start switch will illuminate, and the stop switch illuminator will go out. The tape will run until it is stopped by a prerecorded 1000 Hz cue signal. The sec-cue and ter-cue indicators will momentarily light if and when their recorded signals pass by the playback head. The VU meter(s) on record

Figure 5–32   The Broadcast Electronics series 3000A stereo recorder-reproducer. (Courtesy Broadcast Electronics, Inc.)

models will indicate playback audio level. The playback may be stopped by depressing the stop switch, but the tape will not be recued for future playback. Once the machine has stopped, the cart may be removed by pulling it out of the machine.

To record on the series 3000A, make sure that the machine power switch is on. Insert an erased cart into the deck opening and against the right side. When the cart is in, and properly seated, the stop switch indicator will illuminate and the machine is ready. To start the recording process, depress the record switch between the two VU meters. The record indicator will light. Feed audio signal from the console and adjust the front panel level control(s) on the 3000A for normal levels on the recorder VU meter(s). In the record mode the meter reads the input to the recorder.

Start the tape by momentarily depressing the start switch. The start switch illuminates, the stop illuminator goes out, and a 1000 Hz cue tone burst is automatically recorded on the cue track of the cart tape. Start modulation immediately to ensure a tight cue when the cart is played back. The tape will automatically stop at the stop cue tone burst when the tape loop returns to that spot, so the modulation should be accurately timed to conform to the length of the tape on the cart.

In addition to the 1000 Hz primary stop tone cue, two additional cue tones are available: SEC at 150 Hz and TER at 8 KHz. The sec (secondary) tone is generally used to activate another device at the end of modulation. In series 3000 machines so equipped, this tone is used to activate the fast-forward cuing. The tertiary (ter) tone is user-determined to activate another device or to provide an audible warning that a cart is nearing its end at stations using carts instead of records for music playback. When the sec or ter tone systems are employed, their in-

dicators are illuminated. The auxiliary tones are recorded on the tape cue track when the user presses their front panel switches. As long as the switch is pressed the tone is recorded. When the operator is recording the 150 Hz tone to provide automatic fast forward, the tone must not end before the program material ends. Fast forward can be accomplished manually by pressing the spring-loaded F FWD switch on the front panel. This advances the tape at three times normal speed to the next cue tone.

Broadcast Electronics' 5300C in normal configuration is a three-deck playback-only machine that can be ordered as mono or stereo with or without the secondary and tertiary cue tones (Figure 5-33). Additionally, a separate unit containing a record amplifier can be ordered to make deck number 3 a record-reproduce deck. As with all multideck cart machines, a single motor and capstan drives the tape on all the decks. The cart is placed in the deck opening against the right side, and the stop button is illuminated to indicate that the machine is ready. The user pushes the start button, which will illuminate, and the ready illuminator will go out. The 5300C will accept NAB AA, A, BB, and B carts.

Some recommendations for making good cart recordings on any cart machine are as follows:

1. Always start with a completely erased cart. New from-the-factory carts come with a test tone on them, so they too must be erased (degaussed) before use. Bulk erase a cart on both sides, and then tip it up on its open end and erase again.

2. Locate the splice on the cart tape. It should be visible at the open end of the cart just past the playback head pressure pad. Since cart tape is wound "oxide out," the splice will be on the

Figure 5–33 The Broadcast Electronics 5300C multi deck playback. (Courtesy Broadcast Electronics, Inc.)

"black," or lubricated, side of the tape. The reason for locating the splice is that recording over the splice will cause a "bump" or dropout of modulation during playback. If the splice is visible it ensures that the cart has been recycled and is cued for immediate playback and that the splice has not been recorded over.

3. Clean the heads, capstan, and pinch roller before every recording and as often as possible before playback. Graphite lubricated tape is used in all carts because of how the tape is fed by being pulled from the center of the reel. Some of the lubricant is always deposited on the capstan–pinch roller combination, causing tape slippage and, on the heads, clogged head gaps.

4. When dubbing from any other recording to cart, start the feeding machine first and then start the cart recorder. Often, activating the start switch of the feeding machine will cause a "pop" to be recorded on the cart if the cart tape is started first. Similarly, stopping the feeding machine while the cart is in record mode may record a pop on the cart tape.

5. Before dubbing another tape to cart, equalize it to emphasize its high end. As tapes are copied, each generation of dub loses high frequencies. Too, constant replay of cart tapes causes high-frequency loss due to oxide wear and the pick-up of residual magnetism from playback heads.

SynchroStart

A Henry Engineering product, which helps prevent the pops discussed above, is the SynchroStart (Figure 5–34). A turntable-recorder synchronizer with automatic turntable start-muting, it eliminates "cue burn," or record surface noise and acci-

Figure 5–34 The Henry SynchroStart. (Courtesy Henry Engineering.)

dental record turntable wow-in, and takes the guesswork out of dubbing records to tape cart by muting turntable audio during disk start-up. It starts the cart recorder in record mode at the exact moment that the beginning of disk audio is beneath the playback stylus, and it does all this automatically.

ITC Cartridge Recorder-Reproducers

The International Tapetronics Corporation of Bloomington, IL, a division of 3M, makes the ITC cart machines. The "Delta" series includes the Delta I, a single-cart deck reproducer, the Delta III, a three-cart deck reproducer, and the Delta IV recording amplifier (Figure 5–35). With the addition of the Delta IV recording amplifier, a Delta I reproducer or the bottom deck of a Delta III reproducer can be converted into a recording deck. The reproducer decks accept size A and AA carts.

The front panel features of the Delta IV record amplifier include two VU meters that automatically switch between record (input) and reproduce (output). Centrally between and below the meters is the red record switch. To its left and right are the

secondary and tertiary cue tone record switches. At the bottom, from left to right, are one input level control; the meter select switch, which allows selection of record, play, cue tone, or record bias; a 1 KHz cue tone add-defeat switch; and the second input control.

The Series 99, ITC's top-of-the-line model (Figure 5–36), is available in a reproducer deck and in a record amplifier that, when used in conjunction with the deck, makes the reproduce deck into a recorder. The deck comes with an ELSA function (erase, locate, splice, azimuth). It accepts NAB A and AA size carts and includes high-speed cuing and secondary and tertiary cue tones. The deck has three front-panel illuminated switches: the top switch, which lights green, is start; the middle, which lights yellow, is stop. It lights when a cart is loaded and flashes to indicate that a cart has been played. The lower cue switch lights blue and causes the deck to enter high-speed recue. The deck has three LED indicators below the start switch: green, which indicates detection of 150 Hz cue tones; red, which is the power indicator; and yellow, which indicates 8 KHz tone detection.

The record amplifier has two VU me-

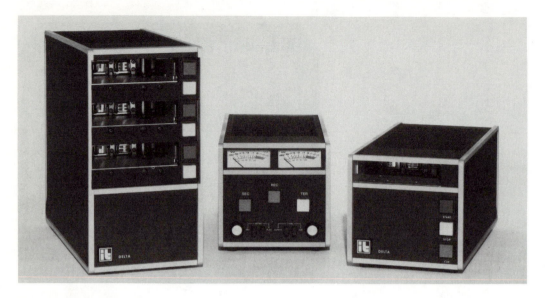

Figure 5–35 The ITC Delta series cart machine. (Courtesy ITC Division of the 3M Corporation.)

ters, each with LED peak indicators, located at the top. Below and between the meters is the illuminated red record switch. Below, to the left and right, are the blue secondary and white tertiary cue tone switches. At the bottom extreme left and right are the level controls. Between these controls is a plate-covered special-functions panel of ten push buttons that provide cue record and defeat, cue erase, and a seven-function test-tone generator.

LINE INPUTS

Line inputs arrive at the console at the relatively high levels of +4 to +8 dBm (or VU), which like inputs from tape playbacks, require no preamplifier in the input chain. If a high-level line-input signal must be fed to a console input with a preamp that cannot be bypassed, it must be "padded" down to a lower input level. Padding is performed by placing a fixed resistive pad, usually from within the console's patch panel, in the input circuit before the preamp.

Line inputs arrive from remote program points, feeding special-event programming to the station, or from the network with which the station is affiliated, feeding programs on a predetermined schedule. All the lines to a station are furnished for a fee from a telephone company, according to a schedule set by the FCC in tariff number 260. The incoming lines may be equalized by the telephone company to provide better frequency response than the usual subscriber-service telephone lines. Lines within a given city or local area are called local loops.

When loops are ordered, they are requested either full time, for continuous use, or part time, for specific time use. They are also ordered either unequalized, or equalized in one of three frequency band designations: a 5 KHz line, an 8 KHz line, or a 15 KHz line. The 5 KHz line is equalized from 100 Hz to 5 KHz; the 8 KHz line, from 50 Hz to 8 KHz; and the 15 KHz line, from 50 Hz to 15 KHz. An unequalized line is less expensive than an equalized line in either full- or part-time

Figure 5–36   The ITC series 99B record amplifier and reproducer. (Courtesy ITC Division of the 3M Corporation.)

service. Incoming lines from distant cities may use the facilities of AT&T Long Lines, an IXC intercity charge based on mileage is added to the cost.

When line quality is not considered of major importance, as in small radio station telephone call-in programs, then stations often feed the output of an ordinary subscriber service telephone line through a hybrid coil and a delay device, to monitor for profanity, to a line input to the console.

REVIEW QUESTIONS

1. Define a console input. Name four types of input.
2. Name three types of broadcast mikes according to directionality. Describe the directional characteristics of each type.
3. What are XLR-3s, and how are they used? What is a strain relief?
4. Why do we use disk reproducers? At what speeds?
5. Describe a tone arm. Describe a phono cartridge.
6. What is a stylus? Describe its tip.
7. What are a TT's quality factors? What is a stylus flat?
8. Enumerate the types of tape playback. Discuss and compare three types (not brands) of tape playback equipment.
9. What is a puller? What speeds does a puller use?
10. What is a multitrack machine? Describe its use.
11. How do we erase a cart?
12. What are the advantages of cart over disk?
13. Why do we find the cart splice before recording?
14. Name the three designations of equalized telephone line by frequency response.

6

CONSOLE OUTPUTS

The reader knows that one of the control console's functions is to route program signal to wherever necessary.

Each program channel of a console terminates in at least one console output. The console output (one output here, for clarity's sake) is fed by wire circuits to wherever its use is required. If that use goes beyond the building housing the control room, then Telco circuits are used as described in Chapter 5.

LINES OUT

At a facility with more than one control room, the outputs of all of the control rooms are fed by lines to a central transmission point, often known as a master control. At this central switching point a master control operator monitors the studio outputs for audio quality and proper levels and switches the program sources on their predetermined time schedules to their destination points. Destination points can be the station's transmitter, the station or recording company's recording room, a broadcast network, or a client's listening room where a salesman might be previewing a program for a prospective sponsor.

A one-studio radio station might have lines out only to its transmitter. A recording company could employ internal lines from its control room console, one line

(one pair) for each program channel, to an input (or track) of its mastering tape recorder.

LOUDSPEAKERS

A console monitoring circuit output, usually of the program on the air, is fed to a monitor buss. This buss then feeds loudspeakers throughout the station, in the station manager's office, the program director's office, the chief engineer's office, employee lounge, sponsor's auditioning room, and the station's reception room.

The loudspeakers used for various rooms are chosen based on reproductive quality needed in a particular area. If the audio is simply background or is used to determine that program is on, medium-quality single speakers in inexpensive enclosures are chosen. In a client's listening room, where fairly critical listening takes place, higher quality combinations of high- and low-frequency speakers, fed by crossover networks and in better enclosures, are chosen. In the control room, where very critical listening occurs, speaker combinations and enclosures that give negligible coloration to the audio are used. In the studio, monitoring is done for several reasons, with fairly good quality speakers and with earphones. When a studio mike is live, people in the studio can only listen with earphones. For in-

Figure 6–1   On-air warning light. (Courtesy Fidelipac Corporation.)

stance, someone reading poetry might want to hear the background music being played. When live music is performed either in the studio or on stage, the performers often have difficulty hearing each other if the music is loud or the audience raucous. And they must hear each other. For the musicians, a monitor feed, called foldback, is made to either their earphones or to loudspeakers with low output, aimed directly at them on stage. Since what is heard in any listening environment is a combination of the speaker(s) and enclosure and the room acoustics, room equalization is often used to balance the equation. Although loudspeaker quality is important to the operator, she has minimal control over them, except for their volume.

WARNING LIGHT LINES

Warning light lines are not console outputs, but because they are operated by relays within the console, they are mentioned here.

To inform passersby that a studio is in use, or on the air, a sign is illuminated, usually in bright red, above the studio entrance door.

When a mike is made live in a studio, another red warning light is illuminated in the studio by the console mike switch to alert program personnel that the studio is "hot."

These illuminated warning systems are operated by relays in the console power supply or the mike input that feed illuminating power to the respective warning devices. Figure 6–1 shows an on-air warning light.

REVIEW QUESTIONS

1. What are console outputs for? Where do they go?
2. What is a program output, a monitor output, and audition output? Are they interchangeable?
3. Describe a monitor buss and a program buss.
4. What is a master control? What is its function?
5. Compare loudspeaker quality with loudspeaker use.
6. Why are warning light systems used?

DISK RECORDING, TAPE RECORDING, AND TAPE EDITING

DISK FORMATS

Disk recordings may be classified into four types: instantaneous, pressings, direct to disk, and digital.

Instantaneous

The instantaneous type can be played back immediately after recording. These disks are cut on a recording lathe; the process involves cutting a continuous spiral groove into an acetate coating on an aluminum base plate. Since the material used is quite soft, play-back is limited to a very few times before audible wear is apparent. These disks are usually employed only for record demonstration purposes or in the mastering process described below.

When a record is cut, the modulation information is impressed into the side-walls of the continuous groove by the recording lathe's cutting stylus. Stereo recording cuts the modulation of one channel into the left wall and the modulation of the other channel into the right wall of the groove.

Pressings

Pressings are made by recording a master tape, transfering the master tape audio signal to an instantaneous acetate master disk, coating the master disk with a carbon spray, and electroplating and factory processing to produce a metal master, a metal mother, and finally a metal stamper. The stampers (dies) are used to press the groove into vinyl wafers, which then become records, or pressings.

Direct to Disk

Direct to disk results in pressings but eliminates the entire master tape sequence of the recording process on the ground that the fewer mechanicoelectric steps between the live music and the recording, the better. In this process, the live music signal is fed directly to the disk recorder. Direct to disk is a very expensive recording system because any mistake made in the music performance or in the disk cutting means starting over again from the beginning, since there is no way to edit.

Digital

Digital recording represents the first fundamental change in the industry's system of duplicating the actual waveform, which is produced by a sound source, either by the mechanical method of cutting grooves in plastic or by the magnetic method of energized tape. Both the mechanical and magnetic systems make an imperfect attempt to create a mirror image, or an analogy, of the original waveform. The imperfection of the analogy is caused in the mechanical case both by the limitation of the groove width, called excursion, before it cuts into an adjacent groove, and by the friction noise caused when two mechanical surfaces, the stylus and the groove, rub against each other.

In magnetic recording, system imperfections are caused by the size and quality of the metallic or metallic oxide particles in the tape emulsion, the physical limitation in head-gap distance (the gap can never be zero), and a problem caused in multitracking. In multitrack tape recording, the greater the number of tracks per tape width, or, more properly, the narrower each track, the lower the magnetic

gain available from each track, and therefore the more gain needed from the track's amplifiers.

As amplifier gain goes up, so does amplifier noise. Thus the analog system of recording is limited in that the original sound is degraded by the very process used to duplicate it. The analog system is further limited by the relatively narrow range of loudness that it can reproduce.

Digital recording abandons the concept of directly duplicating the waveform of the sound source. It is described later in this chapter.

LP Manufacture and Handling

Here, in a series of aging but accurate photographs, courtesy of the recording industry, is how that industry makes vinyl LP records.

The first step takes place in the recording studio (Figure 7–1), where the music is recorded on tape on multitrack machines. After recording is completed, any necessary tape editing takes place in an editing facility under the direction of an operator and tape editor (Figure 7–2).

Figure 7–1   The recording studio.

Figure 7–2   Tape editing.

The next step is the mixdown, which results in a final two-track master tape. The master tape audio signal is then transferred, with the aid of a cutting lathe, to a master lacquer disk. This master lacquer is then shipped to a manufacturing plant, where it is coated with a fine spray of silver (Figure 7–3) to provide a base for subsequent nickel plating.

The nickel plating (Figure 7–4) is accomplished with the silvered master lacquer placed on a spindle and immersed in an electrolytic solution containing pure nickel. The surface of the master is coated to $10/1000$ of an inch. The nickel mold is then separated from the master lacquer, and the result is a metal master, a negative of the master lacquer.

Figure 7–3   Silver coating the laquer master.

Figure 7–4   Nickel coating the master.

The metal master is reimmersed in the electrolytic solution and again coated with nickel. Upon separation (Figure 7–5), there is the metal master and a mold that is once again a positive. The various molds

formed are constantly examined for flaws by inspecting the surface microscopically (Figure 7–6) as well as by play-testing.

The approved mold goes back into the plating process to produce the stamper,

Figure 7–5   Separating the nickel mold.

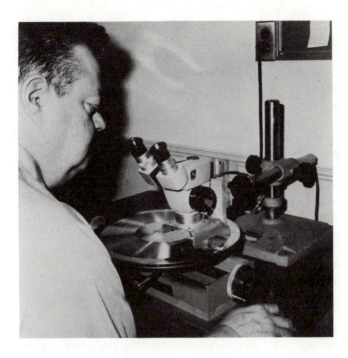

Figure 7–6   Microscopic inspection.

from which the finished commercial record will be pressed.

In another part of the plant, labels are printed (Figure 7–7) on an offset press. After being printed in sheets, the labels are "dinked" to the proper round shape and taken to the record press. The stampers are affixed to the record press; the vinyl compound is blended, heated, and melted; and the labels are placed on the

Figure 7–7   Label printing.

stampers, upper and lower, upside down so that they face outward on the finished record.

The press operator puts the press in operation (Figure 7–8). The compound is placed on the lower label, the press closes, and many tons of hydraulic pressure are applied. At the same time, live steam circulates inside the press, softening the compound and forcing it to conform exactly to the stamper's grooves. The labels are automatically fused to the record during this operation. After a few seconds, cold water circulates through the press, cooling and hardening the record.

After the pressing operation, the record is inspected, packed in its sleeve and album jacket, and shrink-wrapped to protect it until purchase (Figure 7–9).

Handle vinyl LP records only by their edges and label area, never in the groove surface area. This keeps oils and perspiration, which attract and hold dirt and grit particles, out of the groove depressions and protects audio quality.

Always store disks in their inner and outer dust jackets for the same reason, as well as for ease in location and filing. Stack them on end in bins or racks, never lying on top of each other. Storing them horizontally will cause record warp and make playback difficult. Horizontal storage on a hot console guarantees warpage.

Records occasionally become dirty despite precautions. They may be cleaned with a soft cloth, by immersion in warm sudsy water, or by spraying the groove area with a special aerosol cleaner and drying it in a circular motion with a napped cloth.

CD Manufacture and Handling

The CD began as a cooperative effort between two giants in the electronics industry, Philips of The Netherlands, which supplied the optical technology, and Sony of Japan.

The information in, rather than on, a CD is carried beneath its surface in a layer

Figure 7–8   The record press.

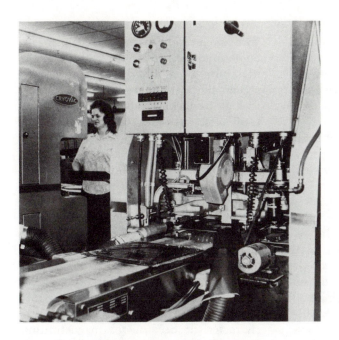

Figure 7–9   Packaging.

that is an optically flat mirror, composed of a thin layer of aluminum upon which microscopic bumps have been raised. In playback, a very small spot of laser light is focused on the information layer, and the bumps affect the way the light is reflected back. Variations in the reflected light are detected and deciphered to read the information on the disk.

Although the information in a CD is normally audio, that is, music for the most part, it can also be, and increasingly is, information input for a computer. In this mode it is known as CD-ROM (random access memory).

The standard CD platter has a transparent base with a diameter of 120 mm. The label side is on top and the readout side on the bottom, with a program area of 33 mm. The readout is optical, so there is no contact and therefore no wear on the disk, and no contact noise is transmitted through the system. The optics of the playback device focus tightly on the information layer, well below the disk surface, so that surface scratches and lightly smudged areas are well out of focus.

The CD manufacturing process somewhat resembles that of the vinyl LP record, except that the mastering part of the process is done under superclean room conditions. The mastered tape, in one method of the cutting process, is fed to a critically accurate glass disk blank, coated with a material known as photo resist. A laser light cutter is focused on the resist coating as the blank revolves. When the information has been fed to it, the blank then undergoes a photodeveloping process to harden the resist in its unexposed areas. The glass blank is then etched to remove the photoexposed areas, creating pits in the resist surface. That surface is then spray-coated with silver. A father is then made by electroplating the blank with nickel. Another electroplating creates a mother, and, from the mother, sons or stampers are produced.

The stampers then are used to stamp plastic disks, thus transferring the structure of the pits to the transparent disk base material. This base material is composed of a polycarbonate plastic with the brand name Makrolon or Plexiglas, which is polymethyl methacrylate.

There are an even number of steps in the process, so that the pits will be raised on the plastic disk surface. This surface is then covered with a thin layer of sprayed aluminum and then by a protective coating of lacquer. The center hole is then punched in the disk, the disk periphery is trimmed if necessary, and the label is affixed to the opposite side. The disk pressing may then be checked optically, boxed, and shipped.

Readout through the bottom thickness of the disk tolerates surface scratches quite well, and severe scratches may be removed with polishing fluids. On the label side, however, the lacquer coating is only 30 μm thick, so that one should do no writing on that side, especially with a ball point pen. The point pressure could distort the information layer and the ink solvents could penetrate the lacquer and reach the information surface.

TAPE RECORDING

Recording tape is manufactured on a base or ribbon of polyester plastic in huge wide rolls or drums, which are later sliced into "pancakes" of the desired width. Widths vary from the ⅛ inch, used in tape cassettes, to the ¼ inch used in cartridge and general-purpose open-reel tape, to the ½, 1, and 2 inch multitracking tape used in open-reel recording. Video recording tapes and digital recorder tapes have still other formats.

The ribbon base is manufactured in several thicknesses, measured in mils, or thousands of an inch. Standard thicknesses, including the emulsion coating, are ¾ mil, 1 mil, and 1½ mil.

The thinner the base stock, the more tape can be wound on a given reel. The more tape on the reel, the longer it can record or play at a given speed. The wider the tape stock, the more tracks and spaces between tracks can be placed on the tape.

The base stock is coated on one side with an emulsion that can be magnetized. Some emulsions are more capable of magnetization than others, and some have specific properties that others do not, such as high-frequency response, but all have trade-offs, such as abrasiveness, which shortens head life. The "standard" emulsion uses an iron oxide as its magnet.

The coating or emulsion is magnetized during the recording process by its contact with the recorder's record head, and magnetization occurs in direct proportion to the audio signal fed to the recording head, provided that the correct high-frequency (well above the audio range) bias current is supplied to the recording head at the same time. Every medium capable of magnetization exhibits a nonlinear characteristic because the magnetization is not directly proportional to the strength of the magnetizing field. This nonlinear characteristic, if not corrected, would severely distort the magnetically recorded material. The use of high-frequency bias current applied through the record head and combined with the audio signal provides the linearity necessary to magnetize the tape in direct proportion to the audio signal. The bias current is supplied to the record head during the record process by the bias oscillator, which is an integral part of the recorder's electronics, at frequencies usually between 100 and 300 KHz.

The base stock for tape used in cartridges is coated on the opposite side from the emulsion with a lubricant to facilitate slippage of the closed loop tape from the center of its reel.

Recording tape is wound on one of three holding devices: open reels, cartridges, and cassettes. Open reels are available in 5, 7, 10½, and 14 inch diameter reels for general-use ¼ inch tape and in larger-sized reels for wider tape. Quarter inch tape is also available on hub-only pancakes for dubbing facilities where the amount of tape on a pancake varies with the amount needed for the dub.

Specifying how long a tape will play or record is a function of the machine's running speed, tape thickness, and size of the tape-holding medium (reel).

Tape cartridge shells are available in standard sizes: A, AA, B, and BB. The cartridge is defined by the NAB as "a plastic or metal enclosure containing an endless loop of lubricated tape, wound on a rotatable hub in such a fashion as to allow continuous motion." Carts use 1 mil tape at a normal speed of 7.5 ips. Some cart machines can operate at optional speeds of 3.75 and 15 ips. The graphite layer of lubricant coating that is added makes the cart tape 1½ mil thick. The endless loop of tape is formed by wrapping the tape with the oxide side out into a spiral. The two ends are spliced together so that, as the tape is pulled from the center of the hub, it passes across the tape heads and winds back onto the outside of the tape spiral.

The cart shell holds, besides the tape on its hub, a braking system and pressure pads that hold the cart tape in contact with the machine's heads. There are three openings across the front of the cart that allow the heads and capstan to penetrate the shell and contact the tape. There is an opening on the bottom of the cart for the pressure roller to rotate through the cart behind the tape. The pressure pad or pads are usually foam, faced with Teflon. These pads must be replaced in time as carts are used. Figure 7–10 depicts a tape cartridge.

Size A carts are purchased with prepackaged tape for 20, 40, 70, 90, 100, and 140 seconds and 2½, 3, 3½, 4, 4½, 5, 5½, 7½, 8, 10, and 10½ minutes. Size AA carts hold tape for 1, 15, 16, and 20 minutes, and size B carts hold tape for 28, 30, 31, 32, 36, 38, and 40 minutes.

Carts can be purchased as shell-only and user-wound for any specific time necessary, within the maximum amount of tape that will fit into the shell.

Since carts are often used repeatedly, they may be stored in the control room on various types of racks like the ones pictured in Figure 7–11. Special identifying labels for prerecorded carts are also available.

Figure 7–10   Fidelipac tape cartridge. (Courtesy Fidelipac Corporation.)

Figure 7–11 Fidelipac cart racks. (Courtesy Fidelipac Corporation.)

Tape cassettes, which are even remotely suitable for broadcast purposes, are made in one standard size. They use ⅛ inch wide tape, and they play and record at 1⅞ ips. Cassettes, like carts, are encapsulated in a shell, but unlike carts employ two internal reels, a feed reel and a take-up reel that stops when the tape has run its course. Cassettes are available in 15, 30, 45, and 60 minutes play-record time per side, since they are usable on both sides.

Because cassettes use a dual- or four-track recording format, they are impossible to edit without first dubbing the recorded material to open-reel tape.

Recording Track Formats

In monaural recording, also called full track and monophonic recording, the signal of one program channel is applied to virtually the full width of the tape (Figure 7–12).

In "half track" recording (Figure 7–13), a monaural signal is fed to the upper half (upper track) of a half track head as the tape travels in one direction, and the same signal is fed to the same head (lower track) as the tape at the end of the reel is turned over, placed back on the feed reel, and then travels in the opposite direction.

Figure 7–12 Monaural recording tape track.

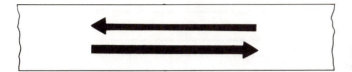

Figure 7–13 Monaural half track.

Four-track monaural recording (Figure 7–14) has a single-program-channel signal applied to one of four tracks in succession, with the tape being reversed at the end of each traverse.

Four-track stereo recording (Figure 7–15) employs signal from two program channels, fed first to tracks one and three; on reversal of the tape at the end of the reel, the two signals are applied to tracks two and four.

In the multitrack formats (Figure 7–16) of 2 (twin), 4, 8, 16, 24, 32, and more track recording, a separate signal is fed from a separate program channel to its separate track as the tape travels in only one direction.

Tape Erasure

Recorded tape may be erased, or degaussed. Open-reel recorders automatically preerase tape with an erase head located immediately before the record head. Cassette recorders use the same erase system as open reel. Most cartridge recorders have no erase facility, so the user must first bulk erase a cart before use. Bulk erasure is a good idea, even with recorders that have erase heads, because residual magnetism often remains on a head-erased tape, causing spurious noise that the much more powerful bulk eraser can remove. Figure 7–17 shows a bulk tape eraser.

The two types of bulk eraser are hand-held, which sits on the tape, and one on which the tape is placed. In either case, the erase method is the same. First, take off your wristwatch and put it a distance away. Put the reel, cart, or cassette of tape on or under the device. Turn the power on. Rotate the tape three or four complete revolutions as the erase magnet hums loudly. Slide the tape out and away from

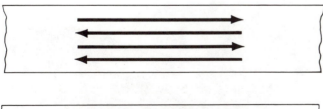

Figure 7–14 Four-track monaural.

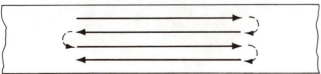

Figure 7–15 Four-track stereo.

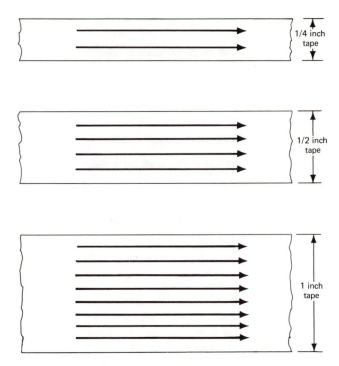

Figure 7–16 Multitrack tape diagrams.

the eraser's magnetic field to arm's length. Turn the power off.

Print-through, or crosstalk between tape layers is the leakage of magnetic energy from tape layer to tape layer. It is most apt to occur in recordings of high-level audio and during the first half hour after recording, when the tape's magnetic field is the strongest. Print-through will also occur if the tape layers are too tightly or unevenly wound, as often happens in re-winding. It can also happen if the recording has been stored for a long time without replay.

Figure 7–17 Fidelipac bulk tape eraser. (Courtesy Fidelipac Corporation.)

TAPE CUT EDITING

A tremendous advantage of tape over disk recording is the user's ability to edit, to cut out a word or phrase or sour note, and then to splice the cut ends together again. Only tapes containing one-track material may be cut edited without destroying material that might be recorded on the other tracks.

Tape may be edited with scissors, provided that the scissors is first demagnetized, but the process is slow and cumbersome. Audio operators employ one of several types of tape-editing devices on the market and mounted conveniently on the tape recorder, usually directly on the head-assembly cover. Cut the tape with a single-edged razor blade or with a razor-edge editing knife. Splice the tape edges together with a specially gummed adhesive tape, manufactured slightly narrower than recording tape, which will not foul the heads of the tape machine and will not seep adhesive and cause adjacent tape layers to adhere to each other.

Two schools of thought embrace tape-cut editing. One school insists on using a yellow grease pencil to mark the tape before cutting and splicing. This system works well, but it also unfortunately coats the playback head of the recorder with residue from the grease pencil, since the tape is marked directly on the playback head.

The second method avoids fouling the playback head and works as follows: Play the tape until the first edit point is reached. Rock the reels back and forth until the precise edit point is reached. Cut the tape at this point and remove the take-up reel, placing it carefully aside. Continue to run the tape through the machine, past the playback head, until the second edit point is reached, allowing the unwanted portion to spill off past the capstan and onto the floor. Rock the feed reel again for the precise edit point and cut the tape at this point. Replace the take-up reel on its spindle, place the two cut tape ends in the splicer, and apply the adhesive splicing tape.

The edit points may be rough if a considerable amount is deleted, and a second edit is performed to tighten up. If the segment of tape to be deleted is saved for later use, run it onto a spare take-up reel instead of onto the floor. Be careful, though, not to mix up the two take-up reels.

If the edit is very short, or tight, save the cut-out piece of tape until the spliced tape has been run through the heads and auditioned. Often even the most experienced editor will cut a syllable or sound too much, and saving the tape scrap until after listening to the edit will save a lot of scrambling around on the floor.

The EdiTall editor (Figure 7–18) is a metal rectangular block with a lipped

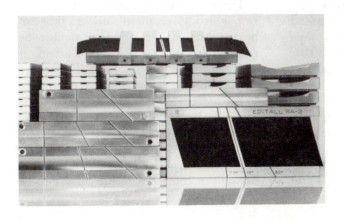

Figure 7–18   EdiTall tape cut editors. (Courtesy XEdit Corporation.)

depression to hold the tape and two cutting grooves. One groove is perpendicular to the tape and the other is at a 45 degree angle. Only the 45 degree angular groove should be used for editing modulated tape to avoid "blips" in the recorded sound. The angular groove effectively provides a sound overlap, or fade, from one tape end to the other rather than an abrupt acoustic change.

To use the editor, place the ends of the tapes to be spliced in the lipped depression, pressing them down smoothly along the entire length with a finger. The emulsion side of the tape should be face down. If cutting is to be performed, overlap the tapes with the edit point (cutting point) directly over the diagonal groove. Make the cut with the blade at a very low angle for a clean cut and then remove the top piece or scrap of excess tape. Place a 1 inch piece of 7/32 inch splicing tape evenly over the joint, so that it does not overlap the edges (width) of the recorded tape. Rub the splice firmly with the index finger to ensure adhesion. Remove the spliced tape from the editor by gently pulling it upward from both ends simultaneously.

LEADER TAPE

Leader tape is a plastic or paper ribbon of the same width as recording tape but in white or some bright color to distinguish it from brown or black recording tape. It serves a number of important purposes. Leader saves valuable recorded material from being torn from the tape end, when that end is whipped from a fast-moving, emptying feed reel in the fast-forward mode. In the rewind mode, leader will save the front end of the tape from damage as it whips from the take-up reel. It saves the recorded tape from bends and foldover as it is attached to the take-up reel.

If leader is joined to recorded tape precisely at the start of modulation, the operator can cue by eye. The tape is threaded until the end of the leader start of the tape is directly over the playback head. The tape is then tightly cued for play without the use of the console audition system.

The titles or timing of individual selections can be written directly on the leader preceding each selection, or the name of the program can be inscribed on the leader at the beginning of the tape for positive identification.

If a pause of specific duration is required in mid-tape, then leader of specific length can be used, 3¾ inches for each second of pause at that speed and 7½ inches of leader for each second at that speed. Bear in mind, though, that when going from recorded tape to leader in mid-recording, there will be an acoustical background sound change.

CARE OF THE TAPE PATH

As the plastic ribbon, coated with magnetic emulsion on one side and often with lubricant on the other, travels along the tape path of tape guides, idler wheels, capstan and pinch roller, and tape heads, it receives two distinct types of "droppings" or deposits. The first is oxide emulsion and lubricant, which coats surfaces, becomes gummy, and moves the tape, if ever so slightly, away from its path surfaces, causing physical distortion of the tape path.

The second deposit is residual magnetism, which builds on all surfaces that can be magnetized. Both deposits must be scrupulously removed as often as necessary. Various guide and head cleaners are marketed and should be used according to the manufacturer's instructions. Heads particularly should be cleaned to remove material from the head gap.

Residual magnetism is removed with a small hand-held degaussing device that is brought in contact with the surface to be demagnetized.

TIME-CODE SYNCHRONIZATION

As multitrack recording began to expand to as many as 60 tracks, a dilemma began in the recording industry: What do we do with the 24-track recorders that were so expensive but were the latest state-of-the-art devices several years ago?

The answer, in some cases, was to get rid of the lesser-track machines and buy the more-track recorders, but the better solution was to add a system to control two or more recorders so that they could be operated "in-sync" and used virtually as one machine.

The system that is employed uses time-code synchronization. It was developed for use in videotape editing at television networks. In its use in audio recording, it links two or more audio tape recorders to a common time index, using time coding on one of the center tape tracks of each machine. This track is, of course, not used for audio but in a sense acts as the sprocket holes do on motion picture film.

There are two codes in general use, both using computer technology, but they are not compatible. One code is called Maglink; the other, pioneered by the Society of Motion Picture and Television Engineers, is called SMPTE code in the United States and EBU code in Europe, after the European Broadcast Union.

The equipment pieces used in a time-code synchronization system are the system controller and the system synchronizer. The controller designates one machine as the master and the other(s) as slaves. The system controller consists of a control electronics rack, a control panel, and a display unit. The control electronics includes the time-code generator, which supplies the code to the tape, and the time-code processor, which includes the time-code reader.

REVIEW QUESTIONS

1. Explain the four classifications of disk recording.
2. How does digital recording differ from analog recording?
3. Describe briefly the step-by-step procedure in the making of an LP pressing.
4. How does one handle a pressing when using it in a program?
5. In what manner should disks be stored? Describe the optimum temperature and humidity conditions in a record library.
6. What is recording tape base ribbon made of? Its emulsion?
7. In what form is audio information stored on tape?
8. Explain bias current in tape recording.
9. How many times may a reel of tape be played? What are the limiting factors?
10. What are the standard tape thicknesses? At what speeds is tape recorded?
11. What size open reels are used? What size carts?
12. Discuss recording track formats.
13. Describe tape erasure. Describe optimum tape storage.
14. Define print-through. How can it be avoided?
15. Describe two methods of tape editing. How is multitrack tape edited?
16. Describe four purposes for leader tape.
17. What is time-code synchronization? Why is it used?

8

PATCHING

The newcomer to the broadcasting or recording industries might well ask questions like these: What happens during a recording session or a broadcast if an amplifier fails, a key goes bad, or any other item of equipment in operation breaks down? Do we stop the program or the recording and repair the faulty item? Do we have spare equipment items to replace the faulty one? How do we connect auxiliary equipment items, like external equalizers, or other effects equipment to our console facilities?

The answers to all these questions and more are found in a system of rapidly bypassing, replacing, or adding equipment items that is called *patching*.

USING THE PATCH CORD AND PATCH PANEL

Patching employs short, very flexible cables with an identical plug at each end and a patch panel into which the control operator inserts these cables. A supply of the cables, called patch cords, in varying lengths usually hangs inside an equipment rack in the control room. The patch panel or patch bay is found sometimes on the face of the equipment rack, sometimes at one end of the console, but always within easy reach of the operator. The patch bay consists of a series of jacks or connector

holes, each labeled on a labeling strip by function (Figure 8–1).

Most operable items of equipment in the control room and many of the items in the console, such as amplifiers, have both their input and outputs appearing on the patch panel. That is, many items (pot, key, amplifier) are connected to both their preceding and following item through the patch panel. When these items are thus connected, undisturbed by the insertion of patch cords, they are said to be normal-through connected. When they are feeding each other through patch cords, they are said to be *patch connected*.

To illustrate patching, we will go back to our basic one-mike input, single-program-channel console. Instead of a direct connection between the mike, preamp, pot, key, program amp and line, as originally depicted, the items are connected in the same sequence but this time are connected through the patch panel (Figure 8–2). This means that the operator can, by inserting a patch cord plug into any input or output on the patch panel, break the connection between any two items and have that input or output available at the loose end of the patch cord (Figure 8–3).

If the operator patches into a preamp input, he has that input at the other end of the cord. Conversely, if he patches into an output, then that output is electrically at the loose end of the cord.

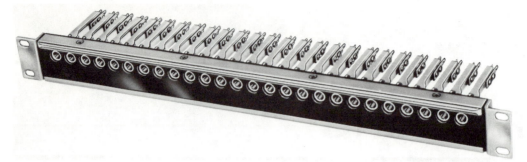

Figure 8–1   Patch panel. (Courtesy Switchcraft, Inc., a Raytheon Company.)

Assume, then, that the preamp in Figure 8–2 fails during the program. Without the patch panel the operator would be faced with the loss of the mike that feeds the system through the preamp. In Figure 8–4, with the patch panel, the operator plugs a cord into the mike output jack, giving him the output of the mike at the cord's loose end. He then plugs that loose end into a spare preamp input jack, thus connecting the mike to another preamplifier. Another cord is then rapidly plugged into the spare preamp output, and its loose end is plugged into the pot of the original mike chain. The faulty preamp is thus replaced electronically by a spare, with the whole operation taking about 15 seconds of an experienced operator's time.

In Figure 8–4 the preamp is the defective member of the microphone chain. Of course, any item could become defective and require replacement, and if a spare of that item is available, it can be replaced if the operator patches around the defective item. Amplifiers tend to be replaced most often. Spare amplifiers are located either within the console or in the nearby equipment rack.

In addition to replacing defective console components, the operator can use patching to connect external equipment to the console or to connect the console output to additional external sources.

To patch connect an equalizer located external to the console, between a mike preamp and the mike pot, we patch the preamp's output jack to the equalizer's input jack and the equalizer's output jack to the pot's input jack.

To connect the console output to other than the line out, we patch the line-out jack to a mult (multiple). A mult on a patch panel consists of three or more jacks on the patchfield, internally connected to nothing but each other, or "floating." When the console output is connected to a mult, all the jacks of that mult become the console output. They are the same point electrically. We then immediately connect another mult jack to the line itself to complete the circuit that was broken when we lifted the line-out jack. That still leaves one or more mult jacks having console output available to be patched elsewhere.

IMPEDANCE MATCHING

Before we patch that console output elsewhere, however, we must consider its impedance. Impedance is a characteristic of every alternating current electronic device. It is abbreviated as Z and is the AC equivalent of resistance. Electronic theory tells us that for a maximum transfer of energy between electronic components, their impedances should match, or be the same in their number of ohms. Ohms is

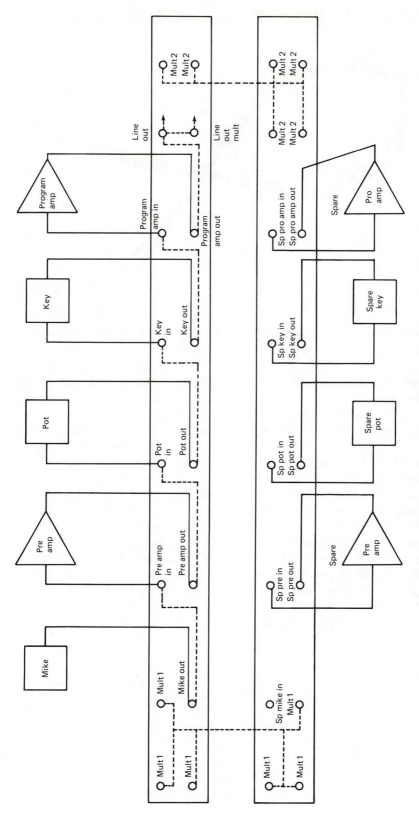

Figure 8–2 Single-mike input channel, normalled through patch panel.

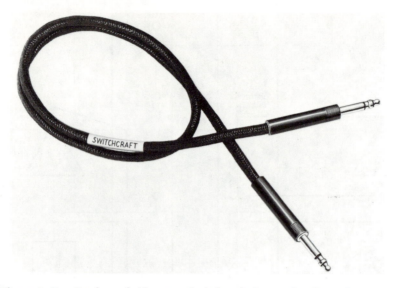

Figure 8–3   Patch cord. (Courtesy Switchcraft, Inc., a Raytheon Company.)

the measurement figure of both impedance and resistance.

When there is an ohms mismatch in a
circuit, there may be severe frequency-
response loss, except if the mismatch is
such that the output of one component
(e.g., 600 ohm) feeds the input of another
component (e.g., more than 10,000 ohm).
That is called *bridging,* where there is no
frequency loss, but there is a signal-level
loss, which must be made up by the gain
of the bridging device or by an amplifier
following the bridging device. Every amplifier has a characteristic input impedance and output impedance.

Preamplifiers are designed so that their

input impedances will match mike output
(150 to 200 ohm).

Preamplifier output impedance is designed to be 600 ohm (remember the definition of the dBm: 1 mW of power into
600 ohm equals 1 VU). This is to match
the inputs of line, cue, and monitor amplifiers, all of whose input and output
impedances are 600 ohm.

After the operator patches the console
line-out jack to a mult, and then patches
from the mult back to the line jack, the
situation is exactly what is was before
patching, except that the console output
is also patched into the mult.

If the operator then patches that mult

Figure 8–4   Defective preamp
replaced with spare.

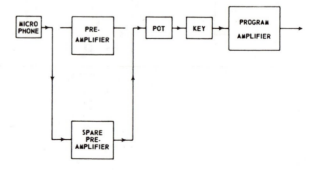

Figure 8–5   Water analogy: half-inch hose to half-inch hose.

to another 600 ohm input, perhaps another line, then the console is "looking" at two 600 ohm impedances in parallel. Electronic theory again tells us that two 600 ohm impedances in parallel are equal to 300 ohm, thus inadvertently halving the console's output impedance and producing a severe mismatch. If we patch the mult to yet another 600 ohm input, the three 600 ohm impedances in parallel make a combined 150 ohm, an even more severe mismatch.

The operator must patch the mult to an impedance combination that matches 600 ohm and that matches the console output. Alternately, he can patch the output to a bridging input. Incorrect impedance connection will "load" the circuit feeding the signal. *Loading* in this instance is defined as overloading, and it is crucial to patch connect each piece of equipment to its recommended load impedance. This is the ohms value that the device should "see" when looking at the circuit that it is driving. That value must always match, or be much larger, to prevent frequency-response loss.

Impedance matching can be viewed with perhaps a bit more clarity by using a water hose analogy. If we connect a half-inch water hose to another hose of the same diameter (Figure 8–5), assuming a constant water pressure, then there is the maximum transfer of water from one hose to the other.

If we connect a half-inch hose to a quarter-inch hose (Figure 8–6), the water flow will be impeded at the connection, and, due to the mismatch in hose sizes, only half the water will flow through the quarter-inch hose.

If we connect a half-inch hose to a 3-inch hose (Figure 8–7), then all the water from the half-inch hose will flow into the 3-inch hose unimpeded, but the water pressure in the 3-inch hose section will drop to one third of its original pressure, since the 3-inch hose could accept much more water than is available from the half-inch hose.

We call this last analogy *bridging*. A bridging input, an input of much higher impedance—10,000 ohm or more—than the output of the component feeding it, will accept the feed with a concomitant signal loss but, in the case of electronic signal, no signal quality loss.

MATRIXING

Another method of component or circuit connecting and disconnecting is called *matrixing*. It employs banks of switches connected to each other internally (similar to a telephone's touchpad) and with

Figure 8–6   Water analogy: half-inch hose to quarter-inch hose.

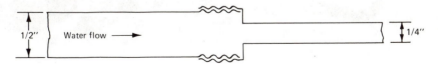

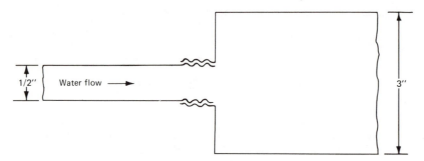

Figure 8–7   Water analogy: half-inch hose to three-inch hose.

touch buttons that illuminate when depressed. Matrixing is used primarily to connect external stand-alone facilities (e.g., equalizers) to a console.

EQUIPMENT RACKS

All the stand-alone equipment in a control room is designed to be 19 inches wide and is mounted with strong bolts in standard-sized 19-inch-across racks. The racks are mounted in cabinets (Figure 8–8).

REVIEW QUESTIONS

1. Why do we patch? What is normal-through connection?
2. What is a mult? Why is it floating?
3. Define bridging. Define loading.
4. Explain the procedure, step by step, for patching around a defective program amplifier.

Figure 8–8   Equipment cabinet racks. (Courtesy Atlas/Soundolier.)

9

LEVELS, BALANCE, AND OPERATING TECHNIQUES

We turn now from the science of audio operation and toward the art of that operation. The operator is faced daily with program situations that require her to apply aesthetic values. Her sense of precise timing and feelings for aural quality can determine a program's artistic excellence. She must have, or must attempt to develop, program sense to be an effective operator.

The operator's actions on her equipment must be deft and accurate. She must have a high degree of manual dexterity, since she stakes her reputation for skill on that dexterity every time she fades in a music selection or keys in a mike.

Did the music come blasting in? Did the announcer sound bigger than the rock group he introduced? Was the mike opened too late to catch his first few words? Was the music clipped out instead of faded?

These questions point up the negative aspects of audio operation. To aid the new operator in a positive approach, we include next some of the principles of good operating practice that are accepted in the broadcast and recording industries and some techniques used by experienced operators.

LEVELS

Levels refer to the degree of sound loudness that is read on the console VU meter and is simultaneously heard on the monitor speaker. Levels are important artistically in that they help determine whether the listener, the ultimate consumer of the program, hears clear or distorted sound, and whether he must strain his ears to hear or is blasted from his seat by excessive sound volume.

Surveys and tests taken many years ago by the Bell Telephone Laboratories, which are still valid, indicate that no matter how loudly or softly an individual operates his receiver, he desires to hear music and speech at the same "relative" sound levels. Further, listeners prefer "even" sound levels regardless of the type of programming and become annoyed at abrupt level changes.

The FCC was deluged at one point with letters from television viewers who complained that commercials were broadcast louder than the program surrounding the commercials.

Abrupt volume-level changes may occur between one speaker and another, between the end of speech and the music that follows, between a loud fanfare opening (music) of a program and subsequent quiet speech, between one music passage and another, between the end of one program and the beginning of the next, with sudden loud sound effects and screams, and with sudden loud audience reaction, applause, or laughter.

The operator should make changes in volume level gradually. He should anticipate loud passages, noises, and sudden level drops, using either experience or rehearsal as a guide.

The range of correct levels as read on the VU meter should vary between -7 and 0 VU for speech and popular music, not higher than -3 VU for audience reaction, no higher than 0 VU on normal music passages, and between -15 and -10 VU on very low passages in symphonic music.

Before the start of a program the operator "takes a level" on the talent in the studio. This is his starting point as far as levels are concerned, and during the program he varies the pot settings as the sound volume changes to stay within the prescribed levels range. The monitor speaker helps in this task when the operator's eyes are away from the meter.

BALANCE

Balance concerns itself with the difference in sound volume between any two or more program components. For example, a firm-voiced interviewer with a hushed-voiced interviewee presents a minor balance problem. Blending the instrument sections of a band or orchestra correctly is a major balance problem.

Balance is a function, to some extent, of room acoustics and types of mikes used as well as of the sound sources themselves. Balance is achieved by adjusting the console pot for each mike used so that the level, as read on the VU meter for one mike, is the same as read for the others. Here, too, the monitor speaker plays a part as the operator listens critically for balance. Once a balance is reached between two or more mikes, the overall volume may be controlled with the master pot.

Do not make your technical operations apparent to the listener. A technical operation may be described as any manual procedure (fading a record in or out, keying a mike in or out) that affects a program.

If a record blasts or drags in at the wrong speed, or "wows" in at increasing speed until it finally sounds normal, or if speech comes in "upcut," with the first few words missing, or ends prematurely in the middle of a word because the key was clipped before speech ended, the technical operations of broadcasting have been made apparent to the listener. She, the listener, then has had her point of focus diverted from the program content to the audio operator.

We have outlined the broad aspects of operating technique. The specifics are covered in the chapters following.

REVIEW QUESTIONS

1. What is a level? What is it measured in? Why are levels important?
2. Describe the correct level range for speech, for pop music, for audience reaction, and for symphonic music.
3. How does the operator vary levels?
4. How do we take a level?
5. Describe sound balance. How is it achieved between two voices?

10

RECORD CUE-UP, RECORD START, AND TAPE PLAYBACK

This chapter discusses how to insert the contents of records, reel-to-reel tape, and tape cartridges into the program format.

RECORD CUE-UP

In its larger sense, a cue is a go-ahead signal in broadcasting. Relating this to recorded material, cuing a record or tape means preparing it on its playback machine so that it is ready to play at the operator's instant demand. The recording is then said to be cued-up.

Before the program begins, the operator should balance the control board. In the balancing procedure, all input pots and keys should be at normal off position. Monitor and master pots should be at their normal open settings. A tone signal is then fed to both the cue system and the program system simultaneously. The cue amp pot and the program channel master pot should both be set to correspond at 0 VU on the meter. The test tone is then removed, and any input that is cued to a specific level on the cue system will read that same level when played on the program channel.

Record cuing includes both holding and releasing a record with the fingers on a spinning turntable. This is best done by placing four fingers lightly on the record's edge, with the thumb resting on the turntable cabinet, steadying the hand.

Cuing Method Number 1

Cuing method number 1 is as follows:

1. Determine the speed at which the record will be played.
2. Adjust the turntable speed control to set the turntable to rotate at the indicated speed.
3. Place the record on the turntable and set the stylus in the first groove.
4. Place the turntable input on the console in the cue position.
5. Rest the fingers of one hand lightly on the edge of the record, retaining it, and activate the turntable power switch. The turntable will then ro-

tate, but the retained record will stay in place.

6. Gently release the record, allowing it to rotate through the unmodulated grooves. Stop the record rotation on the beginning of modulation by gently replacing your fingers on its edge. Turn the power switch off.

7. When the turntable ceases rotation, rock the record to and fro until the exact start of modulation is reached.

8. Rotate the record counterclockwise (backtrack it) manually about one-quarter revolution, and the record is cued.

Alternate Cuing Method

An alternate cuing method, preferred by some because it does not wear the record groove by backtracking, is as follows:

1. Follow items 1 through 4 of the previous procedure.

2. Note a spot or point on the record label that is easily found with the record rotating.

3. Turn the power switch on, with an eye on the locator spot on the label. Observe the number of rotations of the spot until modulation is first heard, including fractions of a rotation.

4. Retain the record with the fingers and place the stylus in the first groove, as close to the locator spot as possible. Release the record, counting revolutions until one-quarter revolution before modulation was heard.

5. Lower your fingers, retaining the record. Turn the power switch off. The record is cued.

The play procedure is different if a record begins with speech than if it starts with music.

RECORD PLAY

Music Start

1. Take a level on the record and cue it up.

2. Return the input key to the program position and the pot to off.

3. Retain the record with the fingers and start the turntable motor. Hold the turntable input pot with the other hand.

4. To play, release the record gently and smoothly open the pot to the indicated level setting.

If the record was cued too tightly, it will wow in, that is, come in off speed and slowly build to normal speed. If it was cued too loosely, there will be a pause before modulation starts.

Speech Start

1. Take a level on the record and cue it up.

2. Leave the input pot set at the indicated level point, and throw the input key to off.

3. Retain the record with the fingers and start the turntable motor. Hold the input key with the other hand.

4. To play, release the record gently and throw the input key to the program position.

Too-tight or too-loose cuing will produce the results discussed above.

Fades

A fade is a potting operation, which at slow or diminished level is called a slow fade and at rapid speed and full level is called a fast fade. It is employed to create artistic effect. Music may be slow faded or sneaked under a performer's voice so that the listener is not immediately aware that the music has started. A fast fade-in might be used for a triumphal fanfare.

A fade-out is the reverse of a fade-in. If a record is to be terminated before conclusion, its pot is slowly closed until modulation is no longer heard. Music records are never terminated by keying out or clipping the music. Further, when the operator fades out a record that includes both instrumental and vocal music, it is considered more tasteful to fade out during the instrumental portion.

Cross-Fades

The cross-fade is a combination fade-in and fade-out between two records. It is performed by sneaking the second record in, while the first is playing, and then tastefully and rapidly fading the second up full and the first rapidly down and out.

Segue

The segue (pronounced seg-way) is a continuous uninterrupted play of two or more records with no live announcement in between. It differs from the cross-fade in that the first record plays to its conclusion before the second is started, on the last note of the first. Many directors today use cross-fade and segue interchangeably, with segue becoming the more archaic term.

Dead Pot

A dead potted record, a record played on a closed pot, is used when a time factor is involved. If the record must end at a specified time, perhaps as the closing theme of a program, then its exact play

time is first ascertained, and it is started by the clock and played on a closed pot until programmatically needed. At that point it is faded up and in, and it concludes precisely at the required clock time. This is aesthetically better than fading out a record in mid-music to end a show.

Backtracking

When a stylus rests in a record groove, it is at about a 45 degree angle, more or less, to the groove, facing toward the front of the phono cartridge. When the record plays normally, the groove advances with the stylus angle. When the record is manually cued in a fast-paced disk jockey (DJ) program, the groove is backtracked against the stylus angle. This procedure begins to cut, or dig into, the soft vinyl groove from the first backtrack. After a number of cue-ups, this reverse angle of the stylus chews up the beginning of the groove so that the start of modulation sounds scratchy and distorted. On a pop music show this is not particularly important, since the DJ will have several copies of the record if it is high on the charts, and, in any case, after a few days or weeks in the limelight, the record will sink into obscurity and perhaps not be played again. On a classical music program, at the other end of the music spectrum, the life of a record may be considerably longer, and thus additional care must be taken to preserve the quality of the start of modulation. It is with these records that the alternate cuing method is preferred. Sometimes, at stations that really care about their "sound," the cuts or selections to be used on-air are transferred to cartridge before airing to solve the backtracking problem.

To exacerbate the backtracking condition, some forms of music, such as rap music, deliberately move a record back and forth in the groove to create a swishing sound that is part of the music. No judgment is offered or even implied here on the quality of rap music, but it does damage both the record and the stylus.

TAPE PLAYBACK TECHNIQUES

Reel-to-reel tape cue-up and play procedure are as follows:

1. Determine the tape speed for playback. If necessary, adjust the tape playback speed control switch.
2. Place the modulated tape on the feed spindle and an empty reel on the take-up spindle of the machine.
3. Thread the tape, oxide side toward the heads, through the guides and idlers of the tape path.
4. Throw the console input key to the cue position and open the pot. Start the machine in the play position of its mode switch.
5. When modulation is heard, take a level on the tape. Rewind to the beginning of modulation and stop the machine.

6. Rock the reels to and fro until the exact start of modulation. The tape is cued.

To play the tape, arrange its input pot and key in a similar manner to record playback, depending on whether the starting modulation is music or speech.

If the tape has been prepared with leader (see Chapter 8), it can be "sight-cued" by rolling tape until the end of the leader appears directly over the playback head.

REVIEW QUESTIONS

1. Define the expression cue.
2. Describe the cue-up and play procedures for a music record and a speech record.
3. Describe wow.
4. What is a fade-in? A fade-out?
5. When do we cross-fade? How does it differ from a segue?
6. What is a dead pot? What is its purpose?
7. Describe the cue-up procedure for reel-to-reel tape.

11

MICROPHONE USE TECHNIQUES

A previous chapter discussed mikes briefly according to type in connection with their use as inputs to a control board. Here we will consider microphone use and the parameters that permit an intelligent choice of mike to perform a particular job. These parameters include the polar patterns or graphs of each type of mike, which describe its directionality; the frequency-response curves for each type, which describe useful frequency range; and what in this book are called the peculiarity, proximity, or differentiative factors of each mike type, which further narrow its applicability to specific job functions.

POLAR PATTERNS

A polar pattern is a concentric circle graph of a mike's directionality in the pickup of sound, marked off at 0 degrees, 90 degrees, and 180 degrees, with the axis of the mike at the center of the graph facing the 0 degree point. The microphone pickup pattern, then, is an overlay on the concentric circles, describing the mike's directionality.

There are four distinct patterns, each having some variation.

Omnidirectional, or All-directional, Pattern

All-directional microphones accept sound with virtually equal facility from any direction. In some omni mikes, however, the pure circular pattern becomes somewhat egg-shaped as sound frequency rises toward the very high end of the spectrum. Figure 11–1 shows an omnidirectional polar pattern.

Bidirectional, or Figure-Eight, Pattern

Bidirectional mikes accept sound with maximum facility at the 0 and 180 degree points, with minimum facility at either of the 90 degree points, and with varying degrees of facility at in-between points of the graph. This pattern varies as the size of the lobes of the figure 8 are changed in relationship to each other. That is, if either the upper or lower lobe is made larger or smaller, then the pattern changes. Figure 11–2 illustrates a bidirectional polar pattern.

Unidirectional, or Cardioid, Pattern

Cardioid mikes accept sound with maximum facility at the 0 degree point, with

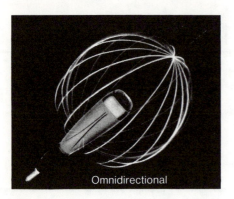

Figure 11–1   Omnidirectional polar pattern. (Courtesy Sennheiser Electronic Corporation.)

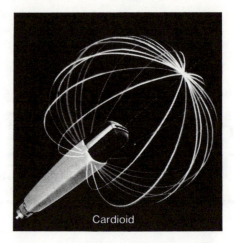

Figure 11–3   Unidirectional polar pattern. (Courtesy Sennheiser Electronic Corporation.)

minimal facility at the 180 degree point, and with varying degrees of facility at in-between points. Figure 11–3 depicts a unidirectional polar pattern. This pattern varies as the size and shape of the lobes below the 90 degree points on the graph are changed. The pattern changes still further, becoming very narrow, as the lower lobes are elongated in the supercardioid and hypercardioid variations. Mikes with these variations are called *shotguns*.

In both bidirectional and unidirectional mikes, the lobe shapes are design characteristics of particular manufacturers, and

in at least two of the older mike types, the lobe shapes can be varied by a recessed switch on the body of the microphone.

Hemispherical Pattern

The hemispherical pattern is found in the boundary or pressure-zone (PZM) mike (Figure 11–4). The PZM picks up equally well in any direction above the surface plane and at all frequencies. A quick look at this pattern shows little to differentiate it from the omni pattern, which is also circular. But adding boundaries adjacent

Figure 11–2   Bidirectional polar pattern. (Courtesy Sennheiser Electronic Corporation.)

Figure 11–4   Hemispheric polar pattern. (Courtesy Crown International, Inc.)

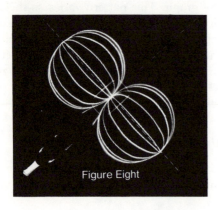

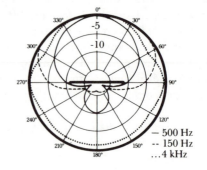

— 500 Hz
-- 150 Hz
…4 kHz

to a PZM shapes the directional pickup pattern. The PZM rejects sound coming from behind the boundaries. Making the PZM directional increases its gain before feedback, and it also picks up a higher ratio of direct-to-reverberant sound. The resulting audio sounds both closer and clearer. Boundaries are described more fully later in this chapter.

FREQUENCY-RESPONSE CURVES

As in polar patterns, the manufacturer supplies a response curve for each microphone as part of its spec sheet. The response curve is plotted on a typical X axis versus Y axis graph, with decibels of gain on the vertical plotted against frequency on the horizontal.

One reads a response curve broadly, looking for the "flat" portion of the curve or for where the response is equal, within a decibel or two, over the greatest range of frequency. We also look at both ends of the curve to ascertain the degree of peaking or, alternatively, roll-off at each end. We look for the working portions of the curve.

The question that this usually evokes from neophytes is, "Why aren't all broadcast mikes designed with a flat curve throughout the audio spectrum?" Primarily because the design would produce a very costly instrument, if it were possible, and also because the manner in which mikes are used requires that some of them have a particular roll-off at one or the other end of the spectrum and that some have a peak at one end of the spectrum.

As an example, lavalier or tie-tack mikes, which are worn on the performer's chest, are below and beneath the performer's chin and therefore must be designed with a particular high-frequency peak and a low-frequency roll-off to compensate for their placement out of the direct line of the sound source.

Both the human ear and microphones respond to changes in the sound wave's pressure component. The amplitude of those variations are measured in microbars, a measurement equivalent to dynes per cm². A sound source at normal listening level at a distance of about 2 feet away is equal to about 1 microbar. Microbar measurement is part of the older centimeter, gram, second (CGS) system that the scientific community has replaced with the MKSA (meter, kilogram, second, ampere) system. The MKSA system uses the newton (N) per m² (N/m²) as its unit of sound pressure. The relationship between the N/m² and the microbar is 1 N/m² equals 10 microbar.

A mike's sensitivity as specified in decibels on its response-curve graph is the alternating voltage in millivolts at the output of the mike, which results when a sound wave with a pressure of one microbar impinges on the transducer element. The figures of sensitivity in decibels are ratios of millivolts per microbar or volts per Newton per meter squared, with the relationship between the two measurement systems as follows:

$$1 \text{ mV/microbar} = 10\frac{\text{mV/m}^2}{\text{N}}$$

For magnetic mikes, the sensitivity figure is quoted with the mike terminated in its characteristic impedance. The sensitivity is measured in an anechoic (echo-free) chamber with the sound wave hitting the transducer on axis or at 0 degrees, unless otherwise stated on the response curve.

To compare the frequency response of on-axis sound impingement with that of sound arriving at the transducer at other angles—that is, to indicate the mike's directional sensitivity, other response curves are often shown on the same graph, with the number of degrees off axis (90, 150, or 180 degrees) stated next to the curve. The difference between response at 0 and

180 degrees, or between the front and the back of the mike, is termed the front-to-back ratio.

Reading a microphone response curve, then, amounts to comparing measured ratios of sensitivity at differing frequencies along the audio spectrum, expressed in decibels. Figure 11–5 shows a typical frequency-response curve.

MICROPHONE PROXIMITY OR PECULIARITY FACTORS

Proximity factors will be described in terms of the idiosyncrasies of particular microphone types.

The dynamic mike, because it uses a plastic or metallic diaphragm attached to an electromagnetic coil, is the most rugged. Dynamic mikes do not require an internal preamp or external (phantom) powering. Although they are subject to the distortive effects of wind noise when used out of doors, a plastic foam wind screen placed over the mike's head will often virtually eliminate the effects of wind blowing across the transducer. Dynamic mikes tend to discriminate in favor of high-frequency sound sources. This factor can be used to advantage when miking a speaker with a particularly deep bass voice to eliminate some of the bass. Because of this sensitivity to the high end of the spectrum, dynamic mikes tend to accentuate the sibilance inherent in some voices. Sibilance can be described as a hissing sound

made in the pronunciation of the letter "S". The typical output level of a dynamic mike is −70 dBv.

True condenser (or capacitor) mikes require a local preamp at the microphone capsule to increase the mike's output gain to a suitable level, typically −40 dBv. Because of the preamp and the resulting higher output level, condenser mikes are said to be "hotter" than their dynamic counterparts. All condensor mikes require external power, either from a local battery or from power fed down the mike cable from the mixer or console, called phantom or simplex power. Phantom power is a DC voltage of 5 to 24 V in Europe and up to 48 V in the United States that is placed across the balanced audio pair at the console preamp and is more reliable than batteries. All condenser mikes operate on similar principles of capacitive action. A charged element, the capacitor, is formed from a stationary back plate and a moving (diaphragm) front plate. Exciting the diaphragm causes a voltage to be produced by the varying capacitance. This voltage is then amplified by the local preamp.

There are two types of condenser mikes. *Full, or true, condensers* use an external power supply both for preamp power and to charge the condenser. This complex circuitry takes considerable space in the body of the mike and explains why it is difficult to make a compact true condenser mike.

The second type of condenser mike is

Figure 11–5   Typical microphone frequency-response curve. (Courtesy Crown International, Inc.)

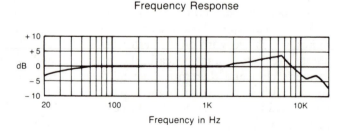

Frequency Response

Frequency in Hz

called an *electret*. Electret mikes feature a permanently charged condenser. Therefore the power supply only operates the preamp; the diaphragm may be tiny, and the circuitry is less complicated and requires less space within the mike body. Small mikes designed to be worn on clothing, the lapel and lavalier mikes, are almost always electrets.

To make an electret, a special high-polymer film is charged with electrostatic energy under high temperature. The polymer film retains this charge almost permanently, even after temperature has returned to normal. The charged film is called electret, and it is stretched across a ring to form a diaphragm that vibrates with sound. As it vibrates, the gap distance between the electret diaphragm and the mike's back plate changes, causing a capacitive change that induces a small variation in the electret film's electrical potential.

The variation in electrical potential is then amplified by the built-in preamplifier and appears as the microphone output signal.

Sony Corporation research has taken the electret one step further and has developed the "back electret," where the electret film is adhered to the mike back plate and causes a thin polyester film to vibrate, which improves sound quality at the low end of the spectrum.

The velocity mike in earlier versions was an extremely fragile instrument, but newer models are quite acceptable for indoor studio use. These mikes are used for performers who have a tendency to "pop." Popping can be described as an overstressing of the explosive consonants—B, P, and T. The explosive quality of these letters causes a sharp momentary rise in the sound wave's pressure component. Since the velocity or "ribbon" mike is actuated by the wave's velocity component, it is less apt to be affected by popping.

Ribbon mikes are most sensitive in the lower range of the audio spectrum. Similar to the single-D cardioid, which will be discussed next, they may be used to advantage to pick up voices that are too high in pitch. Further, the closer that a performer works to a ribbon mike, the lower the pitch will appear to be. This effect is termed *proximity effect*. Velocity mikes are one of the older designs, with current research going in other directions, but they are still produced and used.

The cardioid (heart-shaped pattern) mike appears in two distinct types. The older combination cardioid type was really two microphones in a single encasement. A ribbon transducer and a diaphragm transducer were connected and phased electrically to produce a unidirectional pattern with two or more switchable lobe configurations, and it was alternately switchable to either transducing element alone for an omnidirectional or bidirectional pattern.

The newer cardioid unidirectional mike works—literally—because its case is not sealed. In fact, the sound pressure is invited to contact the mike diaphragm from the back as well as the front, through a port in the back or through slits along one side of the mike. The sound pressure from the rear or side neutralizes diaphragm motion somewhat, delayed by the distance that the sound must travel to the rear or side port and back internally to the diaphragm. The type of cardioid with a port in its back is called a single D, for the single distance from the rear port to the diaphragm. This type has a frequency response that varies greatly with the distance between the source and the mike. At $\frac{1}{8}$ inch from the sound source, the bass response is 15 dB higher than the response if the mike is 2 or more feet away. This proximity effect is used extensively by operators who must add bass to a mike pickup and, alternately, avoided in heavy existent bass sound situations. The Electro-Voice company developed variable D microphones with multiple ports along their side that reduce the bass boosting proximity effect but maintain the unidirectional pattern. Their designs use a variation that they call continuously variable D, where the mid- and low-frequency ports are replaced by a long slotted entrance that has a continuously varying frequency acceptance along its length. Two variations of the cardioid pattern are termed the supercardioid and the hypercardioid. The cardioid has a broad pickup area on axis. Sound from its sides is rejected by 6 dB, and sound from the rear, or 180 degrees off axis, is rejected by 20 to 30 dB. Sound, remember, is doubled or halved by every 3 dB change.

A supercardioid shotgun rejects sound from its sides by 8.7 dB, and it rejects sound best at the two nulls behind the mike, at 125 degrees off axis.

The hypercardioid, also a shotgun mike, has the tightest pattern of the three, with a rejection of 12 dB down its sides, and it rejects sound best at two nulls located 110 degrees off axis.

The hemispherical pattern is found in the "boundary" mike, which is also called the PZM. PZMs take advantage of the acoustic conditions that take place on hard surfaces and particularly at an angular juncture of two or more hard surfaces. A common problem with most microphones is a "comb filter" effect when similar frequencies, with different phases, interact at a mike. The effect can be both apparent and annoying when a multitude of frequencies with random phasing are picked up by a mike in a large concert hall. At the mike, the delayed sound reflections combine with the direct sound, resulting in phase cancellation. What results from this is a series of peaks and dips in the net

frequency response, called the comb filter effect.

It just happens naturally that the reflections from a hard surface will appear coincident (everything in phase) at a short distance away from the reflective surface and thus eliminate the comb filter effect.

In 1978 this acoustic property was put to practical application to create the PZM, which is marketed by Crown International, Elkhart, IN. An electret mike capsule is suspended a very short distance above a flat plate, such that the capsule points downward toward the plate and is suspended within the coincident field. The microphone diaphragm is placed in the pressure zone, just above the boundary where direct and reflected sound combine in phase over the audio range. The hemispherical pattern is thus created because the pickup of the capsule is 180 degrees of an arc over the microphone, in a circumference determined by the physical properties of the surface over which the mike is suspended. The plate, in turn, is placed for pickup at the juncture of two or more large surfaces such as walls or floor and wall. Sound pickup is constant for sound sources at any angle in front of the boundaries and drops off rapidly when the sound source moves behind the boundaries. The bigger the boundary, the lower the frequency at which a PZM mike becomes directional.

The PZM mike has found wide acceptance in all areas of microphone use. A cardioid variation, called the phase-coherent cardioid, has been created. PZMs are relatively inexpensive, have a very low profile and a wide bandwidth, and will sustain very high sound pressure levels (SPL). SPL is a measure of the intensity of sound. The threshold of hearing is 0 dB SPL. Normal conversation at 1 foot measures about 70 dB SPL. Sound that causes pain starts at about 120 dB SPL. PZMs

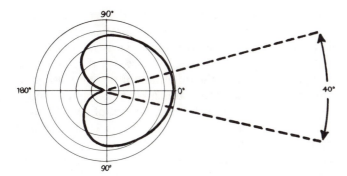

Figure 11–6 Cardioid microphone cone of acceptance.

are also particularly useful for close miking and where mikes must not be obtrusive.

CHOICE OF MICROPHONE

Given the usual choice of either a cardioid or an omni mike, should the operator always opt for the former? Here are the trade-offs: Within the same price range, an omni has a smoother frequency response than a cardioid, is significantly less susceptible to breath noise and mechanical shock, and is often more rugged. It has an all-around pickup, will pick up room acoustics, has extended low-frequency response, and has no bass boost with close use. A unidirectional mike increases the possible working distance between source and mike to a theoretical 1.7 to 1, has a selective pickup pattern, rejects sound from behind the mike, rejects room acoustics, delivers more gain before feedback, and has the proximity effect of up-close bass boost.

Clearly, use determines choice. If use calls for an even, narrower cone of acceptance than the cardioid mike, then supercardioid or hypercardioid line microphones or shotgun mikes are used. Line mikes, such as the Electro-Voice Cardiline, have polar patterns that are much narrower than cardioid mikes; they are

used for pickup at distances beyond even cardioid mike ability, mounted on booms, in television and motion pictures. The polar pattern variations of the supercardioid variable D mike and the Cardiline mike are shown in Figure 11–6. Figure 11–7

Figure 11–7 Supercardioid and hypercardioid polar patterns. (Courtesy Sennheiser Electronic Corporation.)

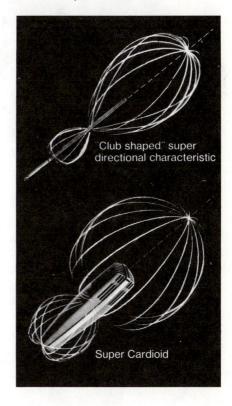

"Club shaped" super directional characteristic

Super Cardioid

shows the supercardioid and hypercardioid patterns.

The directionality of a cardioid mike can be demonstrated by moving a sound source of constant level around the mike at a constant distance from the instrument. Motion of the sound source in any direction away from the mike's specified angle of acceptance lessens the acceptance of sound, and, with the mike face at 0 degrees, a cone of acceptance is described, with the apex of the cone at the transducing element and the base of the cone formed at the sound source. Within the cone area (shown by the dashed lines in Figure 11–6), if the distance between the sound source and mike is increased (the sound source is moved further from the mike), then the base of the cone becomes larger while the angle of acceptance stays the same, but the amount of sound reaching the mike becomes less as the decrease follows the inverse square law.

Figure 11–8 is a cutaway photodiagram of an Electro-Voice RE 18 dynamic cardioid microphone showing the microphone construction. The RE 18 will be described in detail later.

Figure 11–8   Cutaway of the RE18. (Courtesy Electro-Voice, Inc.)

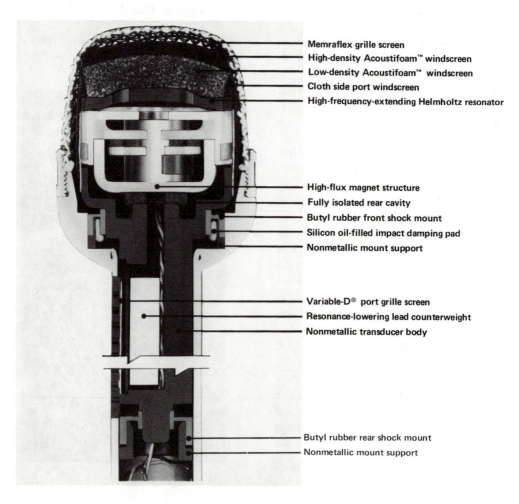

Memraflex grille screen
High-density Acoustifoam™ windscreen
Low-density Acoustifoam™ windscreen
Cloth side port windscreen
High-frequency-extending Helmholtz resonator

High-flux magnet structure
Fully isolated rear cavity
Butyl rubber front shock mount
Silicon oil-filled impact damping pad
Nonmetallic mount support

Variable-D® port grille screen
Resonance-lowering lead counterweight
Nonmetallic transducer body

Butyl rubber rear shock mount
Nonmetallic mount support

Electro-Voice Microphones

We will not show polar patterns for Electro-Voice dynamic omni mikes because they are all identical. The 635A is the most rugged mike manufactured by E-V. It is the one that former E-V vice-president, Lou Burroughs, used to demonstrate its ruggedness by hammering nails into an oak 2 × 4 board, using the mike as the hammer, and then plugging the same mike into its cable and using it in the PA system to continue his demonstration. The 635A has an output of −55 dB, a frequency response of 80 Hz to 13 KHz, and a Z of 150 ohm. Figure 11–9 shows the E-V 635A and its frequency-response curve.

The D056 is an internally shock mounted mike for hand-held use. Han-

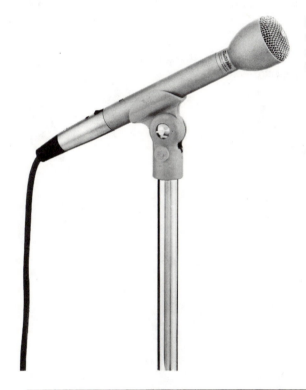

Figure 11–9 The Electro-Voice 635A *(top)* and frequency-response curve *(bottom)*. (Courtesy Electro-Voice, Inc.)

RESPONSE IN DB

FREQUENCY IN CYCLES PER SECOND

dling and cord vibration noise are isolated from the microphone element. It is available in long- and short-handle versions. It is a dynamic omni with a response of from 80 Hz to 18 KHz and an output of −61 dB. Figure 11–10 shows the E-V D056, 56L, and response curve.

The RE55 has the widest frequency response of all E-V dynamic omni mikes, 40 Hz to 20 KHz. The RE55's heritage goes back over 20 years to the original E-V model 655, which was one of the work-

horses of the industry. The RE55 has a relatively flat response, which makes it appropriate where shaped response is not necessary to solve acoustic problems. Its output is −57 dB. Figure 11–11 shows the E-V RE 55 and frequency-response curve.

The E-V DS35 is a single D dynamic cardioid. It has a linear high-frequency response and the typical rise at the bass end of the single D mike in close miked situations. Note the low end difference on the

Figure 11–10 The Electro-Voice DO56L *(left)*, DO56 *(right)*, and frequency-response curve *(bottom)*. (Courtesy Electro-Voice, Inc.)

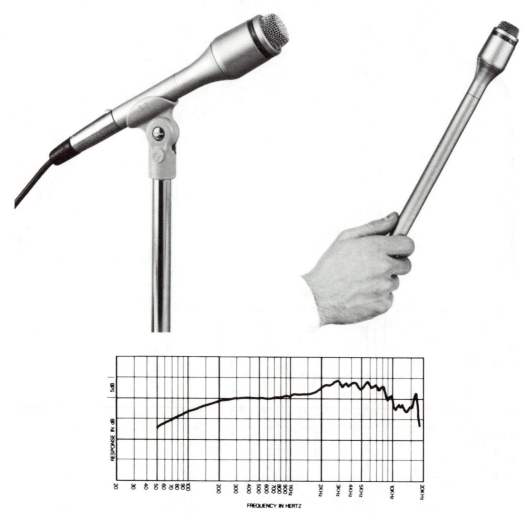

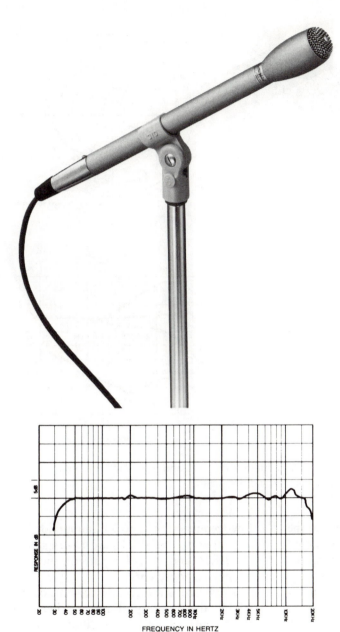

Figure 11–11   The RE55 *(top)* and frequency-response curve *(bottom)*. (Courtesy Electro-Voice, Inc.)

FREQUENCY IN HERTZ

response curve when the DS35 is used at varying distances on axis and also when it is used completely off axis at 180 degrees. It has a frequency response of 60 Hz to 17 KHz and an output of −60 dB. Figure 11–12 shows the DS35 and its polar pattern and frequency-response curve.

The E-V CS15P is a single D true condenser cardioid mike designed to be phantom powered. All true condenser mikes need a source of voltage to operate, and the CS15P gets its source through the mike cable from either the console power supply or from battery supply between the

Figure 11–12 The DS35 *(left)*, polar pattern *(right top)*, and frequency-response curve *(right bottom)*. (Courtesy Electro-Voice, Inc.)

end of the mike cable and the input to the console. The CS15P accepts high sound levels well, making it appropriate for close-miking musical instruments and vocal pickup where extended low-frequency response is desired. Its response is from 40 Hz to 18 KHz, and its output − 45 dB. Figure 11–13 shows the CS15P and its polar pattern and frequency response curve.

The RE15 has great sound rejection from the rear. A variable D dynamic supercardioid, its frequency response of 80 Hz to 15 KHz is unusually independent of the angular location of the sound source. Its roll-off below 150 Hz keeps

low-frequency noise to a minimum. It has an output of − 56 dB. Figure 11–14 shows the RE15 and its polar pattern and frequency-response curve.

The RE16 is a direct descendant of the RE15, the primary difference between them being a built-in blast filter as an integral part of the mike. Note the larger screen at the front of the RE16. It vies with the 635A as the mike most used by audio operators as a general-purpose mike. Like the RE15, it has an internal hum-bucking coil to reduce inductive hum pickup by an extra 25 dB and a bass roll-off switch to eliminate low-frequency noise pickup. Its response is 80 Hz to 15

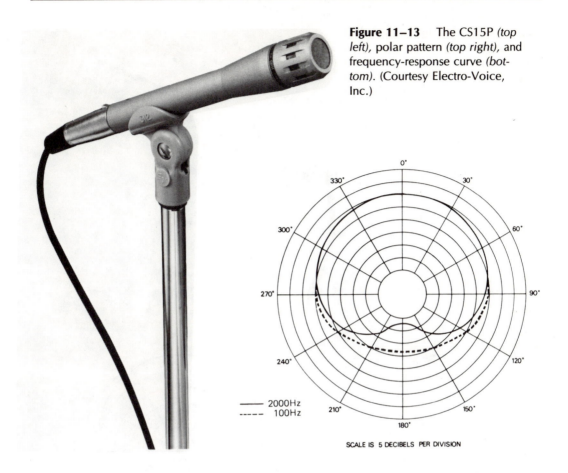

Figure 11–13   The CS15P *(top left)*, polar pattern *(top right)*, and frequency-response curve *(bottom)*. (Courtesy Electro-Voice, Inc.)

—— 2000Hz
- - - - 100Hz

SCALE IS 5 DECIBELS PER DIVISION

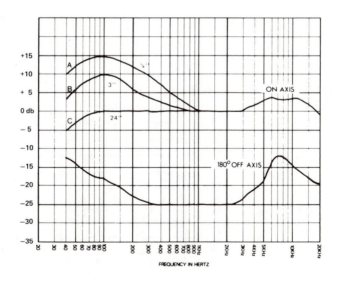

FREQUENCY IN HERTZ

Figure 11–14 The RE15 *(top),* polar pattern *(middle),* and frequency-response curve *(bottom).* (Courtesy Electro-Voice, Inc.)

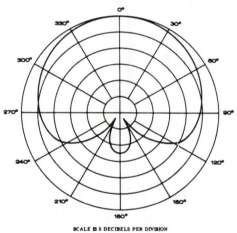

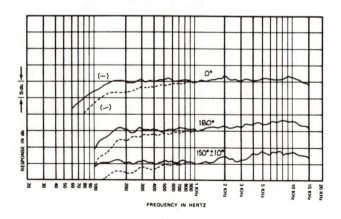

KHz, and its output is −56 dB. Figure 11–15 shows the RE16 and its polar pattern and frequency-response curve.

Another descendant of the RE15, the RE18 (Figure 11–16), has an added internal shock mount for isolating handling and cord noise. It was shown in cutaway in Figure 11–7. It, too, is a variable D dynamic supercardioid, with a response of 80 Hz to 15 KHz and an output of −57 dB.

The RE20 (Figure 11–17) is a variable

Figure 11–15   The RE16 *(top),* polar pattern *(middle),* and frequency-response curve *(bottom).* (Courtesy Electro-Voice, Inc.)

SCALE IS 5 DECIBELS PER DIVISION

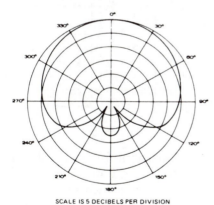

Figure 11–16 The RE18 *(top)*, polar pattern *(bottom left)*, and frequency-response curve *(bottom right)*. (Courtesy Electro-Voice, Inc.)

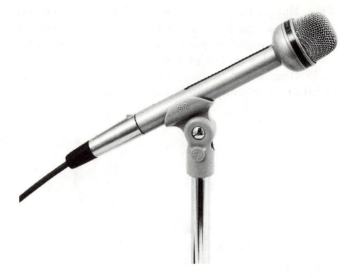

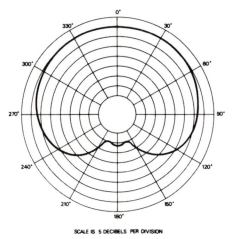

SCALE IS 5 DECIBELS PER DIVISION

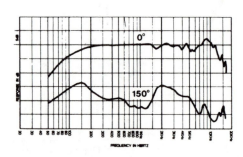

FREQUENCY IN HERTZ

D dynamic cardioid. Its specialized use is for high-sound-pressure level situations (close-miking) where no bass coloration is wanted and when response to the ends of the audio spectrum is needed. It is a relatively heavy mike—1 lb, 10 oz—with a response of 45 Hz to 18 KHz and an output of −57 dB.

The E-V DL42 (Figure 11–18) is a Cardiline dynamic unidirectional microphone. It is 16¾ inches long and weighs 13 oz. One of the shotgun mikes used by news gatherers on a "fishpole" or in TV studios on a shock-mounted boom, it has a response of 50 Hz to 12 KHz and an output of −50 dB. There are diffraction vanes along the length of the line tube to reduce narrowing of the coverage angle at high sound frequencies. It works well at distances of up to four times that of conventional omnidirectional mikes.

The E-V RE45N/D (Figure 11–19) is a Cardiline dynamic shotgun mike 11.5 inches long and 1.87 inches in diameter. It has a response of 150 Hz to 15 KHz when used at a distance and of 50 Hz to 15 KHz when used close up. The polar pattern is that of a cardioid line mike. It

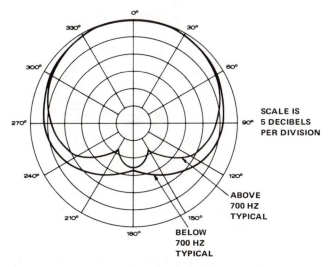

Figure 11–17 The RE20 *(top),* polar pattern *(middle),* and frequency-response curve *(bottom).* (Courtesy Electro-Voice, Inc.)

SCALE IS
5 DECIBELS
PER DIVISION

ABOVE
700 HZ
TYPICAL

BELOW
700 HZ
TYPICAL

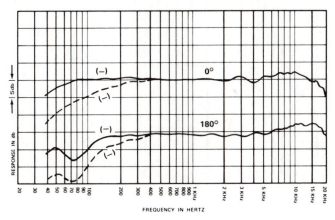

Figure 11–18 The DL42 *(top)*, polar pattern *(middle)*, and frequency-response curve *(bottom)*. (Courtesy Electro-Voice, Inc.)

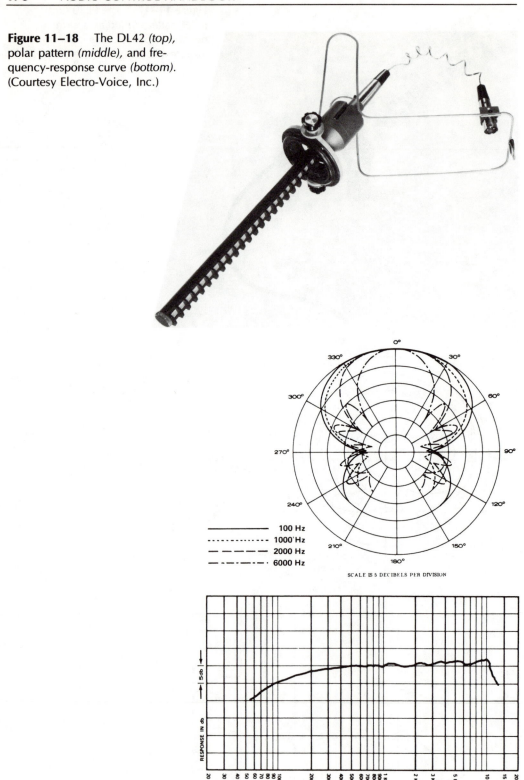

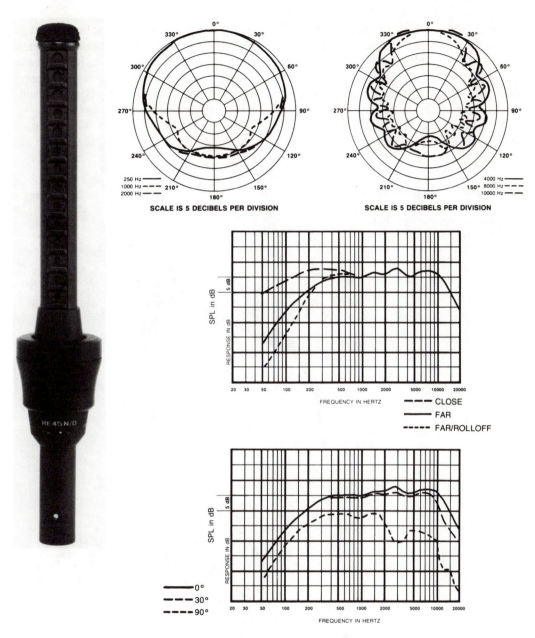

Figure 11–19   The Electro-Voice RE45N/D *(left)*, polar patterns and frequency-response curves *(right)*. (Courtesy Electro-Voice, Inc.)

has smooth off-axis frequency response for a mike that has a distributed front-end design, which is a wave-interference type of line mike.

Shure Brothers Inc. Microphones

The Shure mikes include a polar pattern in the spec sheet for their omnidirectional dynamic mikes, which in every instance indicates that the mikes are designed so that the polar graph becomes egg-shaped instead of round as frequency rises above 4000 Hz and even more elongated as frequency approaches 10 KHz.

The SM61 (Figure 11–20) has a response of 50 Hz to 14 KHz. An omni dynamic with a special flexible grille that protects the mike from damage from a drop of as much as 6 feet, its response range makes it useful for remotes, interviews, sports coverage, or any application where a hand-held mike is necessary.

The Shure SM63 (Figure 11–21) is an omnidirectional dynamic mike with a re-

Figure 11–20 The Shure SM61 *(top),* polar patterns *(middle),* and frequency-response curve *(bottom).* (Courtesy Shure Brothers, Inc.)

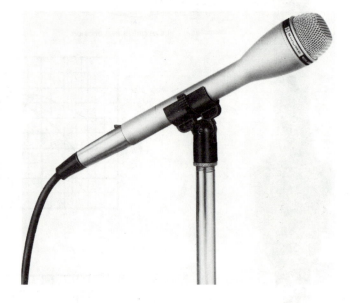

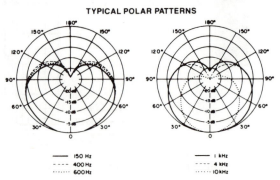

TYPICAL POLAR PATTERNS

——— 150 Hz
- - - - 400 Hz
········· 600 Hz

——— 1 kHz
- - - - 4 kHz
········· 10 kHz

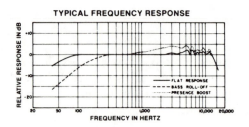

TYPICAL FREQUENCY RESPONSE

FLAT RESPONSE
BASS ROLL-OFF
PRESENCE BOOST

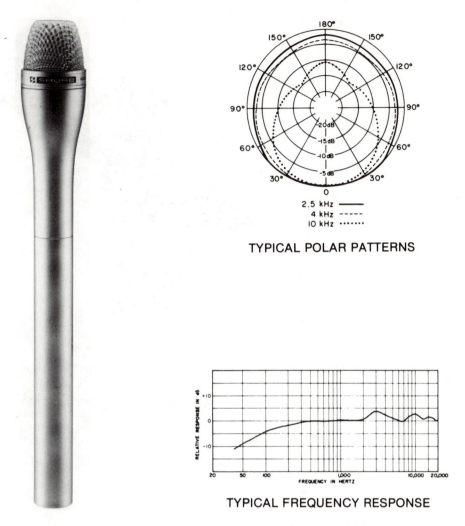

TYPICAL POLAR PATTERNS

2.5 kHz ——
4 kHz - - - -
10 kHz · · · · · ·

TYPICAL FREQUENCY RESPONSE

Figure 11–21   The SM63 *(left)*, polar pattern *(right top)*, and frequency-response curve *(right bottom)*. (Courtesy Shure Brothers, Inc.)

sponse of 50 Hz to 20 KHz. It has a controlled roll-off at the bass end starting at about 300 Hz. Its output is rated to be about 6 dB higher than comparable mikes. It is small and light and has an internal mechanical isolation system. An integral wind and pop filter gives the SM63 low sensitivity to handling and stand noise. Its polar pattern, different for an omni, becomes more "squeezed" with higher frequency.

The SM7 (Figure 11–22) is a unidirectional dynamic microphone designed for the recording studio. It has a smooth, wide frequency response of 40 Hz to 16 KHz, the classic cardioid pattern, and bass roll-off and mid-range emphasis controls (presence boost). It is ruggedly constructed for use on a fishpole or mike boom.

The SM58 (Figure 11–23) is a unidirectional dynamic mike that may be handheld for sports, interviews, and close-to-lips vocals because of its wind and pop filter. It has a cardioid pattern and a response of from 50 Hz to 15 KHz.

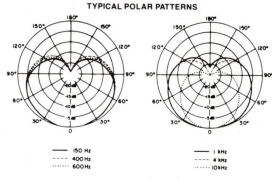

TYPICAL POLAR PATTERNS

—— 150 Hz	—— 1 kHz
- - - 400 Hz	- - - 4 kHz
······ 600 Hz	······ 10 kHz

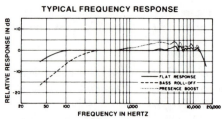

TYPICAL FREQUENCY RESPONSE

Figure 11–22 The SM7 *(top)*, polar patterns *(middle)*, and frequency-response curve *(bottom)*. (Courtesy Shure Brothers, Inc.)

The SM85 (Figure 11–24) is a unidirectional cardioid true condenser mike with a response of 50 Hz to 15 KHz. Because of this wide response it is suitable for use as a hand-held performer mike. It makes use of the single D proximity effect to give the performer control of low-frequency sound. The SM85 uses phantom powering from an external supply, either the Shure Model PS1 or PS1E2, or alternately from the console or mixer into which the microphone feeds.

The SM81 (Figure 11–25) is a unidirectional cardioid true condenser mike

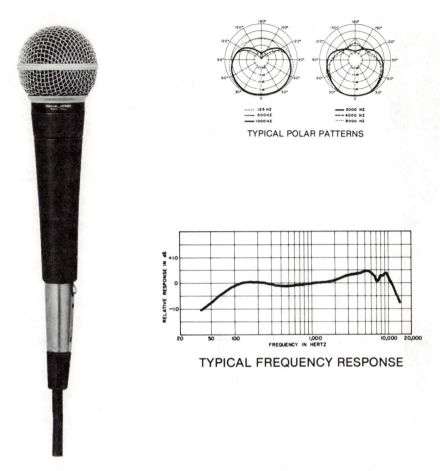

TYPICAL POLAR PATTERNS

TYPICAL FREQUENCY RESPONSE

Figure 11–23   The Shure SM58 *(left)*, polar patterns *(right top),* and frequency-response curve *(right bottom).* (Courtesy Shure Brothers, Inc.)

with very wide response—from 20 Hz to 20 KHz. It is designed for phantom powering. The SM81, because it is a "hot" mike, has a switchable 10 dB attenuator built in. It also has a three-position selector that enables a low-frequency end of the curve, which can be flat or which can have a 6 or 18 dB per octave roll-off. Close-miking with the SM81 causes the proximity effect. Figure 11–26 illustrates this effect in all three switch positions.

Sony Microphones

The ECM 50 was the first small "tie-tack" mike to replace the larger and heavier lavalier mikes. An omnidirectional electret, it has been replaced by a quartet of similar mikes, all of which except the ECM 66 are omnidirectional electrets. The ECM 66 is a unidirectional mike. All can be obtained in either a black satin (B suffix) or nickel satin (S suffix) finish. The ECM 77 is the smallest and the lightest of the four. Its frequency response is from 40 Hz to 20 KHz. Because of its small, inconspicuous nature, it enables unobtrusive miking and minimizes glare. It is the least susceptible mike of the group to rustling noise. Figure 11–27 shows the ECM 77, 66, 55, and 44. Figure 11–28 shows the ECM 77.

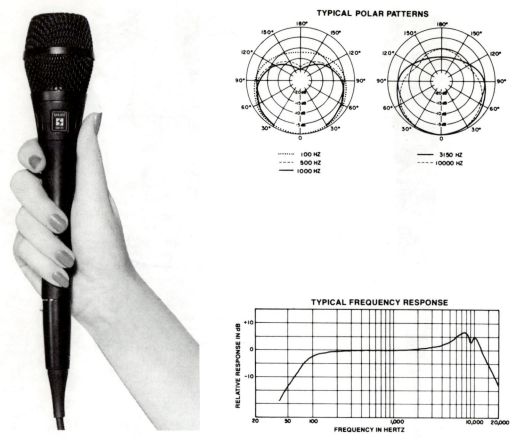

Figure 11–24   The SM85 *(left)*, polar patterns *(right top)*, and frequency-response curve *(right bottom)*. (Courtesy Shure Brothers, Inc.)

The ECM 66 (Figure 11–29) is used primarily for contact miking of musical instruments. Its unidirectional pattern is specifically contoured for feedback control. The ECM 55 and the ECM 44 are less expensive versions. Sony also makes a line of holders for its lavalier–tie-tack mikes. There are single and double mike tie or lapel clips, pencil-type clips for a man's jacket breast pocket, safety pin clips, and necklace-type clips. There are also metal or urethane wind screens in six colors and a belt clip for the mike's power supply. All the mikes may be powered by the included battery supply, and all but

the ECM 44 can be phantom or simplex powered.

The Sony C 535P and C 536 (Figure 11–30) are true condenser unidirectional mikes designed for multimike use. They provide tight, clean, balanced reproduction of musical instruments. The C 535 is designed to pick up sound from the source at which it is aimed. The C 536 is most sensitive to a sound source at a right angle from the mike axis.

The C 74 and C 76 mikes (Figure 11–31) are very similar except for length. The C 74 is shorter by 10 inches, making it more maneuverable in tight spaces. They

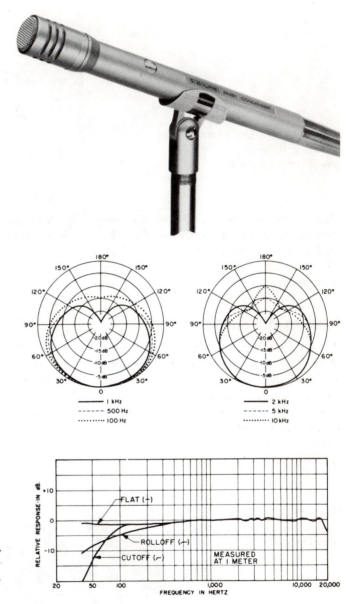

Figure 11–25 The SM81 *(top),* polar patterns, *(middle),* and frequency-response curve *(bottom).* (Courtesy Shure Brothers, Inc.)

are cardioid true condenser omni shotguns with foam wind screens. Their frequency response is 40 Hz to 16 KHz, and they can be either battery or simplex powered. They are each 1 inch in diameter, with the length of the C 74 being $16\frac{7}{8}$ inches and the C 76 $26\frac{3}{4}$ inches. The C 74 weighs 12.6 oz and the C 76 weighs

14.7 oz. There is a pistol grip accessory for each mike.

The C 48 (Figure 11–32) is a true condenser mike with uni-omni-bidirectional characteristics. The mike has two diaphragms, switchable to the required characteristic. An LED indicator describes which position the mike is in. A switch-

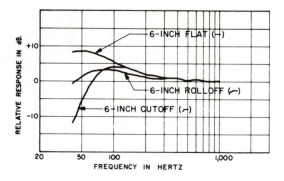

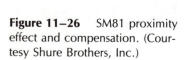
Figure 11–26 SM81 proximity effect and compensation. (Courtesy Shure Brothers, Inc.)

able 10 dB pad and a two-position low-cut switch are part of the C 48. The frequency response is 30 Hz to 16 KHz. The mike can use either battery or simplex power.

Neumann Microphones

These microphones, made by Georg Neumann in Germany, are available exclusively in the United States through the Gotham Audio Corporation. Neumann mikes are the ultimate in microphone quality. Their cost mirrors their quality. I have used them for music recording for the better part of 30 years.

An i suffix on a Neumann tells the operator that the mike uses an XLR-3 connector. They are also available with the European Amphenol-Tuchel connectors.

The U 89i (Figure 11–33) has a pressure-gradient true condenser transducer and a polar pattern that can be omni, wide-angle cardioid, cardioid, hypercardioid, or figure-eight by a switch located below the grille. This makes the U 89i adaptable to about every pickup variation. A fragile instrument, it is subject to audio dropout under high humidity. It has a response of 40 Hz to 18 KHz and can use either battery or phantom powering.

Figure 11–27 The Sony ECM 77, 66, 55, and 44. (Courtesy Sony Corporation.)

(Actual Size)

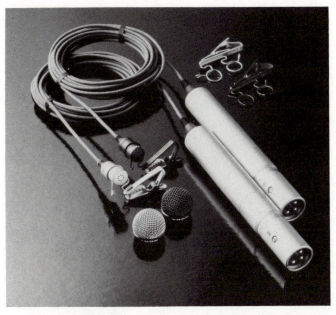

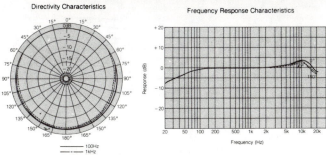

Figure 11–28   The Sony ECM 77 *(top)*, polar pattern and frequency-response curve *(bottom)*. (Courtesy Sony Corporation.)

The family of mikes to which the U 87Ai (Figure 11–34) belongs has been in regular use in studios for more than 25 years. It is the standard music mike in most studios for vocal solos, piano, strings, and brass instruments. A DC converter has replaced batteries in this true condenser. It has switchable pattern characteristics and a frequency response of 40 Hz to 16 KHz.

The USM69 (Figure 11–35) is a stereo true condenser microphone pair, with two mike systems consisting of two transducer elements that can be directionally oriented by rotating the transducers against each other through an arc of up to 270 degrees. The upper windscreen is rotated around the lower to change the transducer orientation. The transducers are the pressure gradient type, and the polar patterns of omni, wide-angle cardioid, cardioid, hypercardioid, and figure-eight are selected separately for each mike system by color-coded left and right switches in the mike body. Additionally, the USM69 can be used to drive two separate mono mike channels if two independent mikes are required at the same precise location. The USM69 has a five-pin DIN connector that can be cable connected to two XLR-

Figure 11–29 The Sony ECM 66 *(top)*, polar pattern and frequency-response curve *(bottom).* (Courtesy Sony Corporation.)

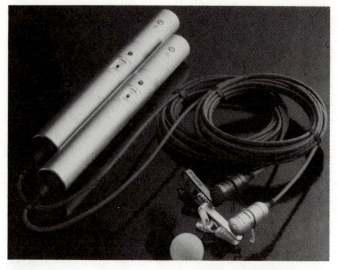

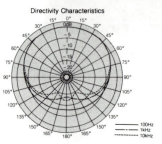

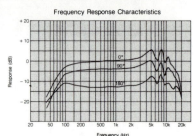

3s. Its frequency response is 40 Hz to 16 KHz, and it requires a battery supply or phantom powering.

The Neumann KMR 82i (Figure 11–36) is a true condenser shotgun mike almost 16 inches long. It has the typical supercardioid lobe-shaped polar pattern and an interference pressure gradient transducer. It weighs 250 g and has a response of 40 Hz to 20 KHz. It may be battery or phantom powered.

Getting the most in directionality in cardioid and supercardioid mikes is limited for physical reasons. To further improve directionality, an interference tube is attached in front of the mike's diaphragm. This tube has a large number of sound inlets distributed over the length of the tube. Each inlet is damped in a specific manner, and this arrangement par-

tially cancels the sound within the tube, depending on the angle of sound incidence.

Sennheiser Microphones

The MD 421 (Figure 11–37) is a cardioid dynamic unidirectional mike with a frequency response of 30 Hz to 17 KHz. It has five positions of low-end roll-off, controlled from a ring-type switch at the base of the mike.

The MD 431 (Figure 11–38) is a popular vocalist mike. It has a response of 40 Hz to 16 KHz and is a supercardioid pressure gradient mike.

The MD 441 (Figure 11–39) is a dynamic supercardioid with a wide response of 30 Hz to 20 KHz. Sennheiser rates this

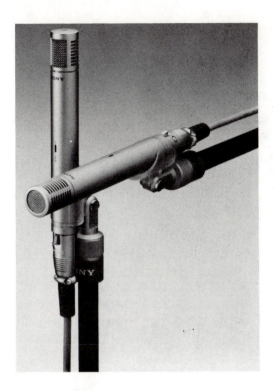

Figure 11–30   The Sony C-535P and C-536 *(top)*, C-535P polar pattern and frequency-response curve *(middle)*, and C-536P polar pattern and frequency-response curve *(bottom)*. (Courtesy Sony Corporation.)

Directivity Characteristics

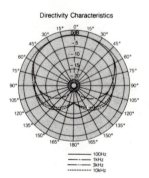

Frequency Response Characteristics

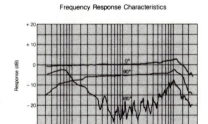

C-536P

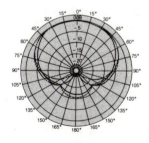

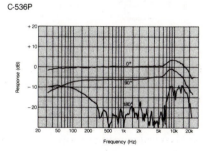

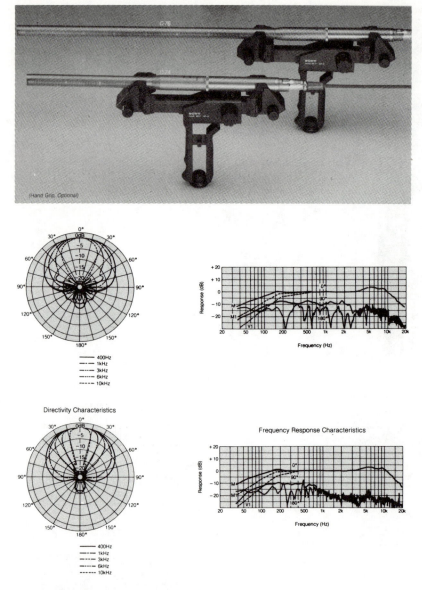

Figure 11–31 The Sony C-74 and C-76 *(top)*, C-74 polar pattern and frequency-response curve *(middle)*, and C-76 polar pattern and frequency-response curve *(bottom)*. (Courtesy Sony Corporation.)

mike as its best dynamic unidirectional. The MD 441 has 10 switchable response curves that adjust response at both ends of the curve.

The MKH 816 (Figure 11–40) is a very popular ENG shotgun mike with a club-shaped polar pattern and a response of 40 Hz to 20 KHz. The 816 has a pressure gradient interference transducer and is powered by an external supply. It is 19 mm (less than 1 inch) in diameter and 55 mm (22.2 inches) long. The 816 weighs 375 g.

The Telemike modular system (Figure

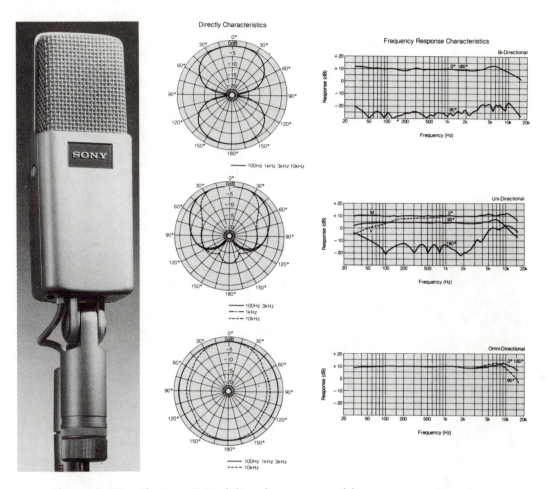

Figure 11–32 The Sony C-48 *(left)*, polar patterns and frequency-response curves *(right)*. (Courtesy Sony Corporation.)

11–41) has as its rationale the idea that in the same way that a professional still camera will accept wide-angle, telephoto, and zoom lenses, so should a sound acceptance system have a variety of pickup heads. The system starts with the K3U powering module, which contains a 600 hour battery to power the electret "heads" and a three-position low-end roll-off switch. Connected to the powering module are the operator's choice of the ME 200 omni head, the ME 40 supercardioid head, or the ME 80 supercardioid shotgun head.

Crown International PZM Microphones

The polar patterns of the Crown International microphones are not shown because they are a perfect hemisphere until placement.

The PZM-6R (Figure 11–42) is an electret, mounted on a 2.5 inch by 3 inch plate. When suspended over an orchestra on a clear plastic panel, the 6R practically disappears. It has a rising high-frequency response, which makes it useful where crisp music attack is desired, such as on percussion instruments, drums, or piano.

Figure 11–33 The Neumann U 89i *(left)*, polar patterns *(right)* and frequency-response curves *(opposite)*. (Courtesy Gotham Audio Corporation.)

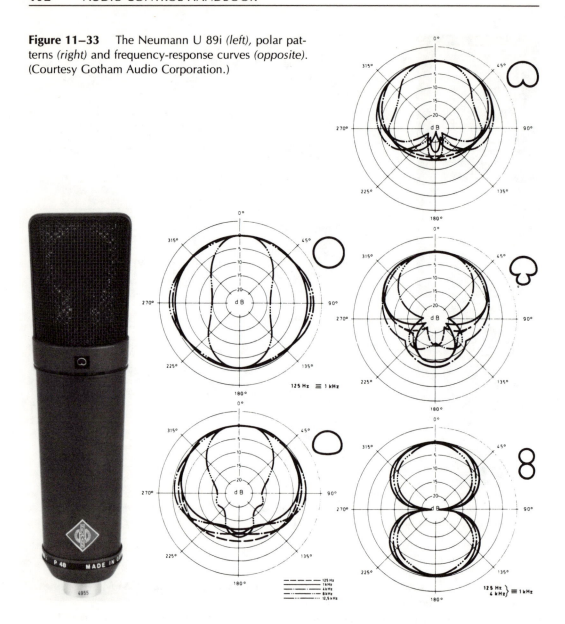

The 6R can withstand up to 150 dB SPL without distortion.

The PZM-20RG (Figure 11–43) is designed for permanent flush mounting in conference-type tables, such as the anchor table of a news set.

The microphone mounts into a standard 4 × 4 electrical box. It may be phantom powered. Avoid mounting it where items such as papers might be placed on top of it. For small conference tables, seating eight, a 20RG should be mounted in the center of the table. For longer tables, there should be a 20RG in the middle of every four to six people. Ideally, no person should be more than 3 feet from the microphone.

The PZM-30FS (Figure 11–44) has a smooth, flat high-frequency response. It is designed to operate on any stiff, nonab-

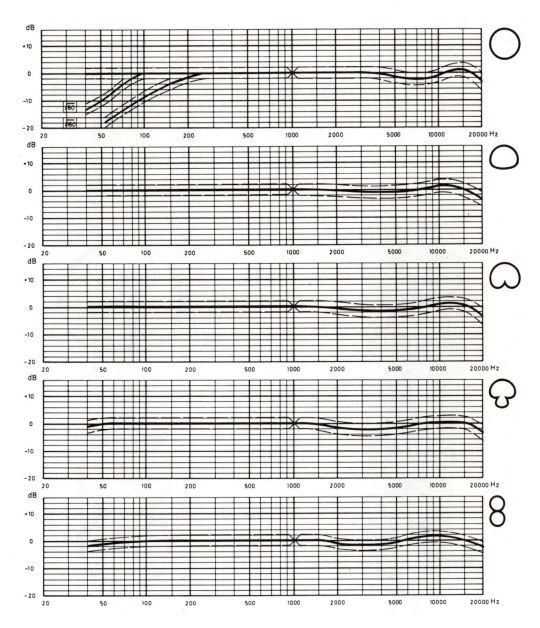

sorbent boundary or surface. Typical boundaries are a floor, wall, ceiling, or table.

Aesthetic Considerations

On-Mike and Off-Mike

A performer should be on mike, or within the polar pattern live area, for normal pickup. Off-mike refers to the so-called dead sides or areas around velocity and cardioid mikes where there are nulls. These areas are, of course, not totally dead to sound, but the pickup will have a far-away, down-in-the-barrel quality. This off-mike effect is sometimes deliberately used to simulate aural distance between two performers, one on-mike and the other off-mike.

Figure 11–34   The U 87Ai *(left)*, polar patterns *(right)* and frequency-response curves *(opposite)*. (Courtesy Gotham Audio Corporation.)

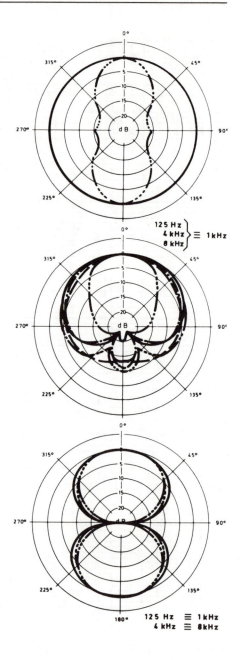

Presence

A performer's presence may be described as his being on-mike at *his* proper distance from the mike. If the performer is too close, then lip smacking, teeth clicking (particularly dentures), tongue slap-

ping, and breathing noises will be heard. The performer is then said to have too much presence. If, however, the performer is too far from the mike, his voice will sound roomy, hollow, and lackluster. He is then said to have no presence.

The performer should be *his* proper

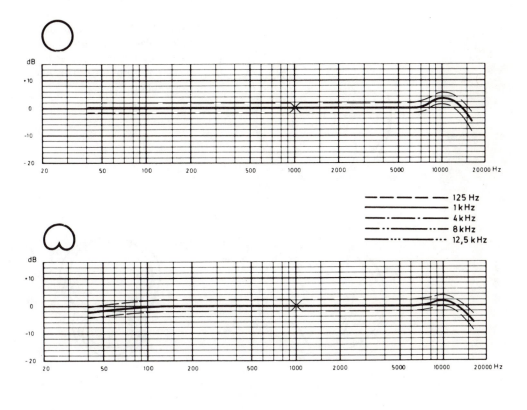

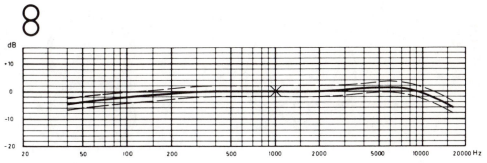

distance from the mike for optimum presence, and that distance will vary from a few inches to perhaps 3 feet away depending on the performer's voice quality. If the audio operator detects either a lack of or too much presence, she signals the performer to change his distance from the microphone.

Clipping a Mike

The operator always clips or cuts the key controlling a mike as soon as its use is terminated to prevent feeding the program channel with extraneous sound such as throat clearing. If the talent calls for a clip by hand signal to clear his throat or

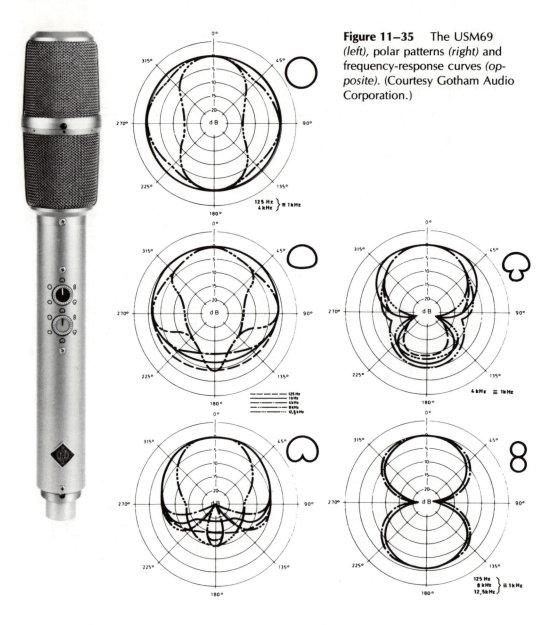

Figure 11–35   The USM69 *(left)*, polar patterns *(right)* and frequency-response curves *(opposite)*. (Courtesy Gotham Audio Corporation.)

to cough, then the operator clips the mike on signal and with the same motion throws the mike key to the cue position, to hear the end of the cough, before returning the key to the program position, also on cue from the performer.

Microphone Phase Reversal

If two mikes are live in the same studio and are feeding the same mixer buss from the same sound signal source, they must be in the same signal phase. If they are in opposing phase, or out of phase, their combined signals would negate each other rather than add. Recall that each mike has an internal diaphragm that reacts to sound impinging on it by creating a minute signal flow of alternating current. If two mikes have one mike cable whose high- and low-side wiring were oppositely wired from the other mike cable, then the signal currents flowing from the mikes into the

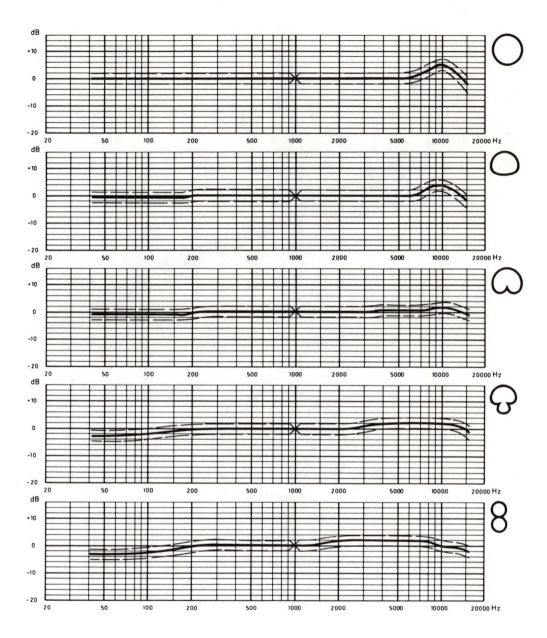

mixer would be out of phase. The condition can be corrected by rewiring the reversed polarity cable or, on some consoles, by throwing a phase-reversal switch on an input module.

Multiple Mike Interference

Frequency-response loss can occur when two mikes of proper phasing are placed too close together. They should be placed so that there is a 3:1 ratio between mike and mike and mike and user. That is, the mikes should be placed so they are three times as far apart from each other as either is to the user. When two mikes must be placed close together, as on a podium, then multiple interference can be avoided by placing the mike heads directly together but not touching each other. If the mikes are cardioids, the 3:1 ratio can be

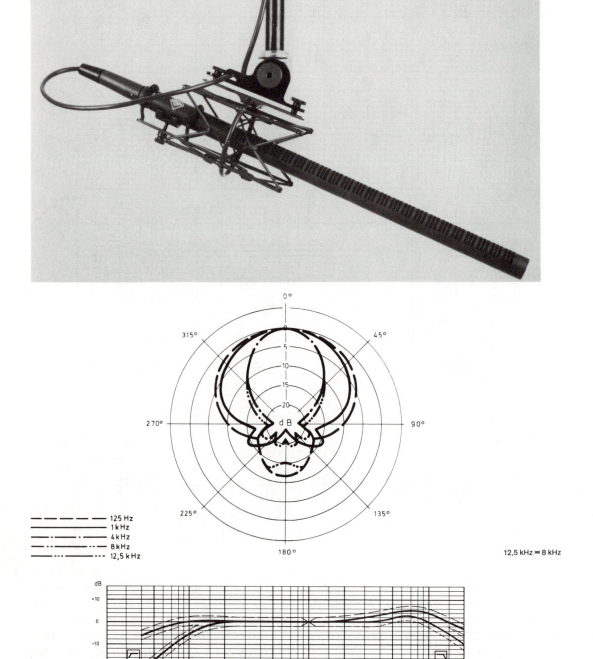

Figure 11–36 The KMR 82i *(top)*, polar pattern *(middle)*, and frequency-response curve *(bottom)*. (Courtesy Gotham Audio Corporation.)

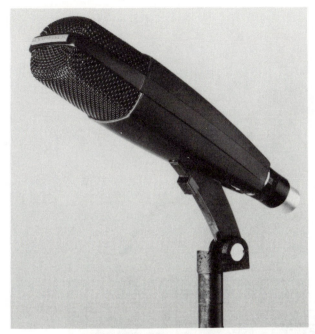

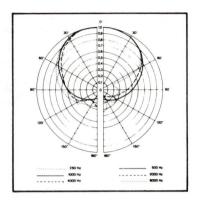

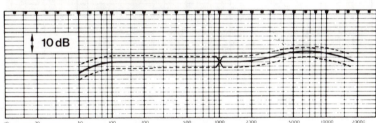

Figure 11–37   The Sennheiser MD 421 *(top left)*, polar pattern *(top right)*, and frequency-response curve *(bottom)*. (Courtesy Sennheiser Electronic Corporation.)

reduced somewhat by angling the mikes away from each other. This microphone ratio was researched by Lou Burroughs, author of the excellent *Microphones: Design and Application,* Sagamore Publishing Co.

WIRELESS MICROPHONE SYSTEMS

Wireless mike systems use virtually every type of microphone, so what we are discussing here is really not microphones but transmission-reception systems to use with microphones instead of microphone cables. The wireless systems have become very popular in broadcasting, where it is often difficult or impossible to run mike cables. Examples are the news reporter on a crowded political convention floor attempting an interview with a candidate; a performer on stage, singing while gyrating or rapidly moving across the stage; the performer singing on a barge anchored 50 feet off shore, with the audience on shore; the speaker at the commencement address, 500 feet away from where the operator and his equip-

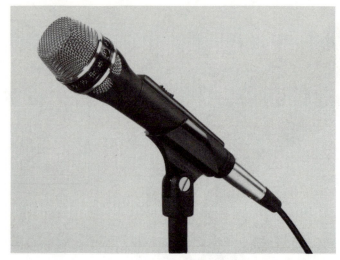

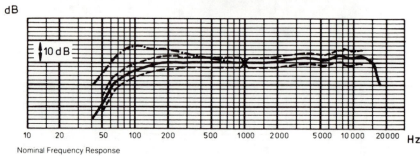

Nominal Frequency Response

Figure 11–38   The MD 431 *(top)* and frequency-response curve *(bottom)*. (Courtesy Sennheiser Electronic Corporation.)

ment must be located. The solution to difficult or impossible cable runs is to use wireless mike systems. And the reason that wireless mike systems do not eliminate mike cables entirely is their cost. They are expensive.

Each mike feeds into a miniature battery-operated transmitter, either integral to the microphone case or externally in a small, flat case that is carried in a pocket or clipped to the belt. The transmitter is very low powered, about 50 mW, for short-range transmission. The transmission range varies from about 250 to 1350 feet. The transmitter frequency allocations are in the 150 MHz to the 216 MHz bands of the UHF FM spectrum.

The transmissions are received by receivers on the same frequencies, often with complex diversity antenna systems to maximize reception. The audio is fed from the receiver to the console or remote mixer. Some systems include two transmission frequencies, the second to be used if the first is not free of interference.

Telex Wireless Systems

Telex makes three systems, built around its FMR-2, FMR-4, and FMR-50 receivers. Each system may use a number of different transmitters or microphones with built-in transmitters.

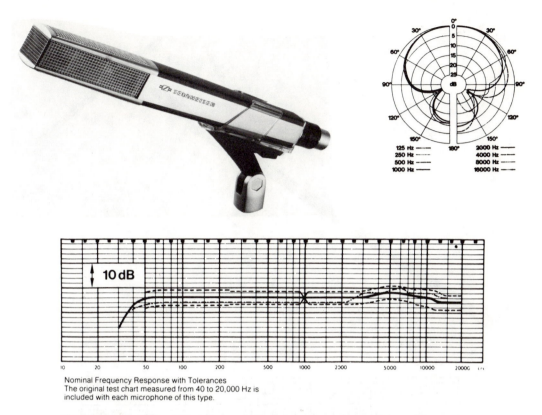

Nominal Frequency Response with Tolerances
The original test chart measured from 40 to 20,000 Hz is
included with each microphone of this type.

Figure 11–39   The MD 441 *(top left),* polar pattern *(top right),* and frequency-response curve *(bottom).* (Courtesy Sennheiser Electronic Corporation.)

The FMR-2 (Figure 11–45) is a single-channel, diversity-antenna receiver. A diversity-antenna receiver employs two antennas, separately phased, and continuously compares the phase relationship between them. It then continuously chooses between antennas for the best signal. The receiver processes the signal and delivers the demodulated audio to the mixer input. Signal to the receiver antennas is sent by one of a number of transmitters.

The WT 50 (Figure 11–46) is a lightweight (5.4 oz) belt pack transmitter, about the size of a cigarette pack, permitting easy concealment. It is powered from a 9 V alkaline battery that provides 8 hours of operation. It uses any one of several Telex electret mikes or the mike built into a Telex headset.

The WHM 500 (Figure 11–47) is a combination electret condenser mike and built-in 15 mW transmitter with a line of sight range of 1000 feet. It is operated with two 4.5 V alkaline batteries.

The Telex WHM 410 is a dynamic mike whose other characteristics are similar to the WHM 500.

The FMR 4 receiver (Figure 11–48) is a rack-mounted four-channel, diversity-antenna, receiver operating in the 165 to 216 MHz range. The FMR 50 is a single-channel, non-diversity-antenna receiver operating in the 150 to 186 MHz range.

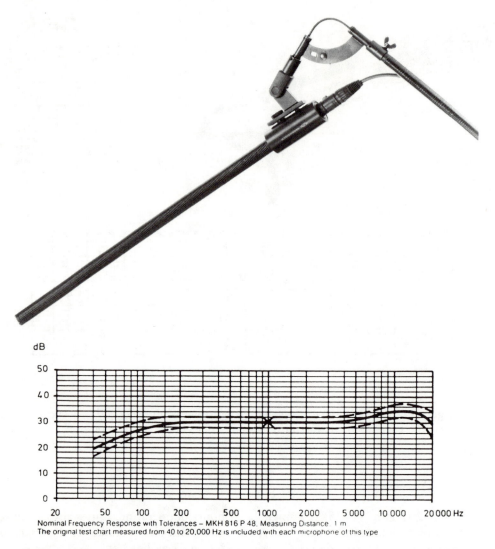

dB

Nominal Frequency Response with Tolerances – MKH 816 P 48. Measuring Distance 1 m
The original test chart measured from 40 to 20,000 Hz is included with each microphone of this type

Figure 11–40   The MKH 816 *(top)* and frequency-response curve *(bottom)*. (Courtesy Sennheiser Electronic Corporation.)

The Telex WT 400 transmitter (Figure 11–49) is a 5.5 oz belt pack transmitter with a transmission range of 500 to 1320 feet, line of sight. It is powered by a 9 V alkaline battery and has a flexible wire antenna. It uses any of the Telex electret mikes and operates on either of two selectable frequencies.

The Telex HT 400 (Figure 11–50) is a mike with built-in transmitter that can operate on one of two selectable frequencies. It is powered by a 9 V alkaline battery, and it permits the choice of three separate mike elements: the Telex TE 10 condenser, the Shure SM58 dynamic, and the Shure SM87 condenser. Shure mikes have been described earlier in this chapter.

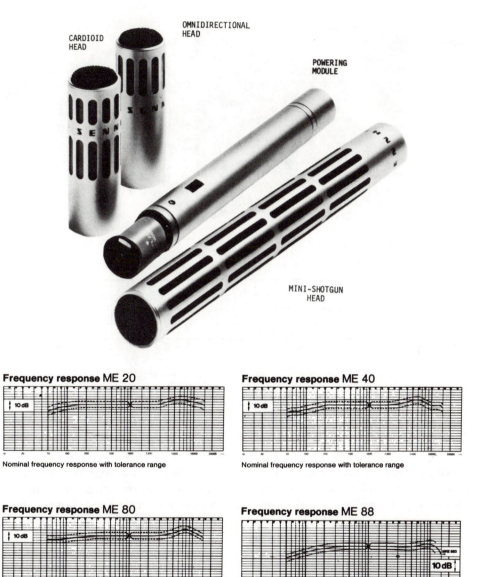

Figure 11–41   The Sennheiser Telemike modular system *(top)* and frequency-response curves *(bottom)*. (Courtesy Sennheiser Electronic Corporation.)

HM Electronics Wireless Mike Systems

HME makes two systems, the system 50 (Figure 11–51) and the system 55 (Figure 11–52). They both employ the same diversity-antenna receiver, the RX520, which operates on either of two channels in the 169 to 174 and the 174.6 to 216 MHz band. The system 50 has a belt pack and a two-selectable frequency transmitter whose XLR-3 connector accepts any microphone input. The transmitter is powered by a 9 V alkaline battery that has a typical life of 6 to 8 hours. The

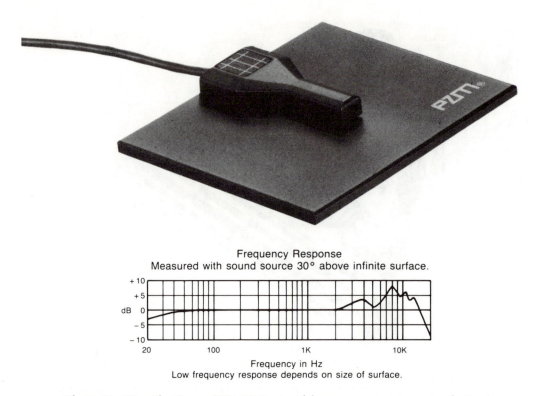

Frequency Response
Measured with sound source 30° above infinite surface.

Frequency in Hz
Low frequency response depends on size of surface.

Figure 11–42 The Crown PZM-6R *(top)* and frequency-response curve *(bottom)*. (Courtesy Crown International, Inc.)

transmitter weighs 4.7 oz and is 4.5 inch × 2.5 inch × .82 inch in size.

The system 55 employs a mike with built-in transmitter. The SM55 mike has a choice of four mike heads or elements: the HME 58, a dynamic, and Shure models SM58, SM85, and SM87. Its transmitter operates on one of two selectable frequencies in the 169 to 174 and 174.6 to 216 MHz frequency bands. It is powered by a 9 V alkaline battery.

Cetec Vega Wireless Mike Systems

Cetec Vega makes the Pro-Plus line of wireless systems, which include both diversity and nondiversity types. Cetec Vega's receivers include the R-33 (Figure 11–53), a miniature, portable, 9 V battery-operated receiver available in both black and cream color. The battery will operate the receiver for about 10 hours if the display, an LED audio-signal-battery readout, is turned off. It operates nondiversity, in the 150 to 174 and 174 to 216 MHz FM bands, with a working range of up to 1000 feet line of site under ideal conditions.

The Cetec Vega model R-42 (Figure 11–54) is a diversity receiver in the 150 to 216 MHz FM band that is available as nondiversity model R-41, with single antenna module. It may be powered from either 115 or 230 V sources, switch selectable, or from an external 15 to 20 V battery supply and at either 50 or 60 Hz power, making its operation useful almost anywhere in the world. It weighs 8 lb 8 oz and has a working range of about

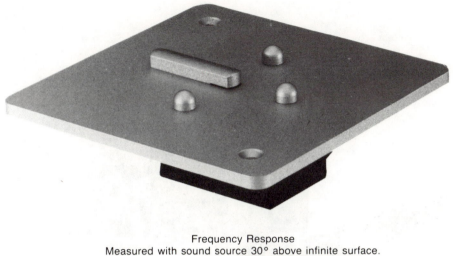

Frequency Response
Measured with sound source 30° above infinite surface.

Figure 11–43   The Crown PZM-2ORG *(top)* and frequency-response curve *(bottom)*. (Courtesy Crown International, Inc.)

Figure 11–44   The Crown PZM-3OFS *(top)*, polar pattern and frequency-response curve *(bottom)*. (Courtesy Crown International, Inc.)

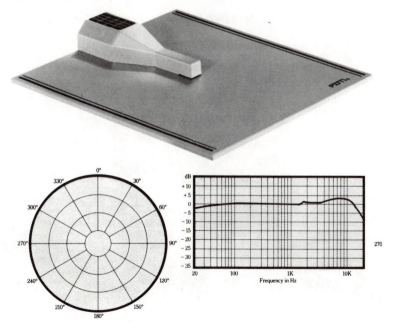

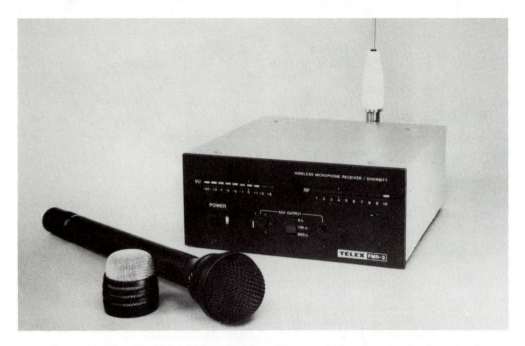

Figure 11–45 The Telex FMR-2 receiver. (Courtesy Telex Communications, Inc.)

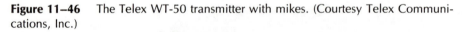

Figure 11–46 The Telex WT-50 transmitter with mikes. (Courtesy Telex Communications, Inc.)

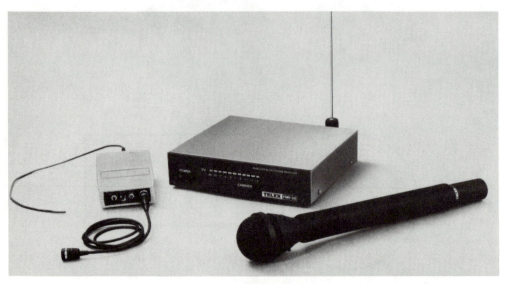

WHM-500 includes an alternate
windscreen for quick styling
changes.

WHM-500   WHM-410

Figure 11–47   The Telex WHM-500 *(left)* and WHM-410 *(right)* microphones.
(Courtesy Telex Communications, Inc.)

1500 feet line of sight under ideal conditions. The front panel has a large RF-VU meter, LEDs to indicate the diversity-selected channel, squelch, the presence of RF signal, and modulation. Controls on the front panel are power on-off, meter function, diversity select, and monitor out jack. On the rear panel (not shown) are audio output XLR-3, audio phase, audio attenuator, 115/230 V select, and two antenna connectors.

The Cetec Vega model 66B (Figure 11–55) is a full-sized portable nondiversity receiver, available also as the 67B diversity receiver. The 66B operates in the 150 to 174 and 174 to 216 MHz FM bands with a working range of up to 1200 feet line of sight. It weighs 2.1 lb and can be powered by either an external 10 V DC supply or by four internal 9 V batteries. Its controls include a meter function switch for audio-signal-battery, monitor

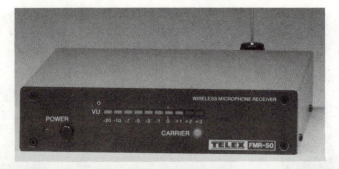

Figure 11–48   The Telex FMR-50 *(top)* and FMR-4 *(bottom)* receivers. (Courtesy Telex Communications, Inc.)

Figure 11–49   The Telex WT-400 transmitter. (Courtesy Telex Communications, Inc.)

Figure 11–50   The Telex HT 400 multichannel mike transmitter. (Courtesy Telex Communications, Inc.)

Figure 11–51   The HME System 50. (Courtesy HM Electronics, Inc.)

Figure 11–52   The HME System 55. (Courtesy HM Electronics, Inc.)

output-level control, audio output switch, mike-line, mike level, and power (internal, off, external). Connectors include XLR-3 audio output, tip-sleeve phone jack, and a BNC coaxial antenna jack. The meter is multifunctional, with a VU scale.

Cetec Vega makes five one-piece mike transmitters (Figure 11–56) with mike head elements by three suppliers: Shure, the SM58 dynamic, the SM85 condenser,

and the SM87 condenser; AKG, the C 535 condenser; and Beyer, the M 500 ribbon. They all (except the SM58) use an internal 9 V battery and an internal dipole antenna printed on the PC board, and they are operable up to 1500 feet under ideal conditions in the 150 to 174 and 174 to 216 MHz FM bands

Cetec Vega also has a model 77 pocket-sized transmitter to which any mike can

Figure 11–53   The Cetec Vega R-33 receiver. (Courtesy Cetec Vega Division of the Cetec Corporation.)

Figure 11–54   The Cetec Vega R-42 receiver. (Courtesy Cetec Vega Division of the Cetec Corporation.)

be attached that is similar in operating conditions to the mike transmitters just described.

Sony Wireless Systems

The Sony wireless mike systems can operate in the 900 or 470 MHz UHF-FM band or a combination of both bands to maximize performance. The systems use multiband operation to permit the use of many wireless mikes during the same performance. Within an assigned frequency band, only a limited number of mikes can be used without causing interference problems. For instance, only five mikes can be used on a performance in the United States in the 900 MHz band and 12 in the 470 MHz band. In Canada, eight mikes can be used on a performance in the 900 MHz band.

The WRT 27 transmitter (Figure 11–57) comes in two models: the WRT 27A for the 900 MHz band and the WRT 27 for the 470 MHz band. The mike used with the transmitter is optional, but it is connected to the transmitter with a cable that has a compact connector known as an SMC on the transmitter end.

The WRT 27 transmitter, a pocket model (Figure 11–57), has 30 mW of output RF (radio frequency) power. It has a $\frac{1}{4}$ wave antenna that is normally mounted

Figure 11–55   The Cetec Vega 66B receiver. (Courtesy Cetec Vega Division of the Cetec Corporation.)

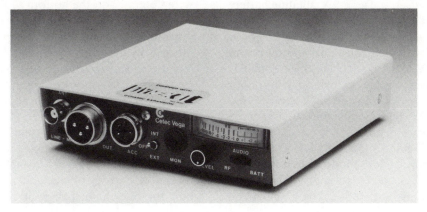

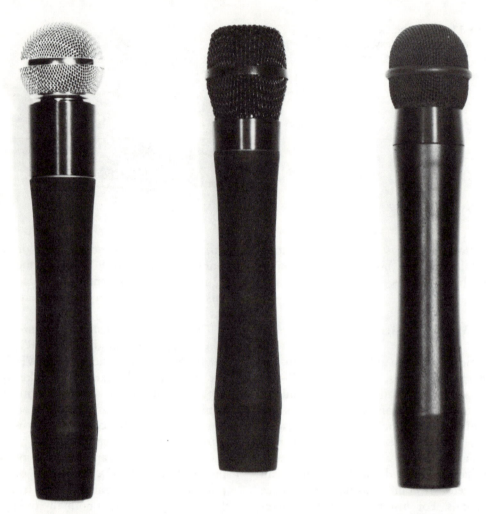

Figure 11–56   The Cetec Vega mike transmitters. (Courtesy Cetec Vega Division of the Cetec Corporation.)

on a headset. It uses a 9 V alkaline battery with an operating life of about 2 hours. It can feed its RF output to a WP 27 power amplifier, which can boost the output to 500 mW. Also available is an external AD 27 power supply to extend continuous operation. Its transmissions are received by either a ground plane antenna or a parabolic antenna.

The WRT 57 (Figure 11–58) is a 900 MHz band microphone and transmitter in one package. It is designed to be used indoors, as in a studio. The antenna is a

wire "pigtail" on the bottom of the mike. The mike is a back electret condenser unidirectional whose transmitter operates at 30 mW in the UHF-FM band. The mike has a frequency response of from 70 Hz to 15 KHz. It is powered with a 9.45 V mercury cell with an operating life of about 2 hours.

Tuner receivers are small units that may be mounted in portable base units. The PB 53 holds three tuners for single-channel reception or two tuners and a diversity unit for one-channel diversity reception.

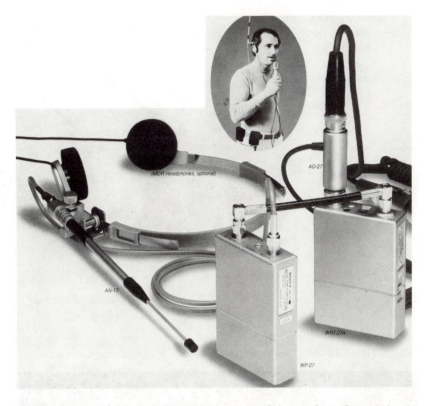

Figure 11–57   The Sony WRT-27 transmitter. (Courtesy Sony Corporation.)

Plugging the tuner or diversity unit into the base provides antenna, output, and power connections. The MB 52 accommodates a total of six tuners or four tuners and two diversity units. Several MB 52s may be mounted in a standard 19 inch rack. Antennas are available for both frequency bands and in single antenna and diversity models. Figure 11–59 shows a WRR 57 tuner receiver. Figure 11–60 shows antennas.

Shown in Figure 11–61 are six typical system configurations for Sony wireless mike combinations.

MIKES IN TELEVISION

On a television program, microphones may be permitted to be seen or may not. This decision depends on the type of program being performed (panel talk show, yes; 18th-century drawing room, no).

Figure 11–58   The Sony WRT-57 mike transmitter. (Courtesy Sony Corporation.)

Figure 11–59 The Sony WRR-57 tuner receiver. (Courtesy Sony Corporation.)

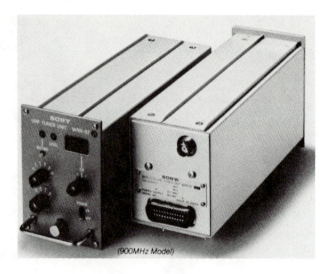

(900MHz Model)

Figure 11–60 Sony antennas. (Courtesy Sony Corporation.)

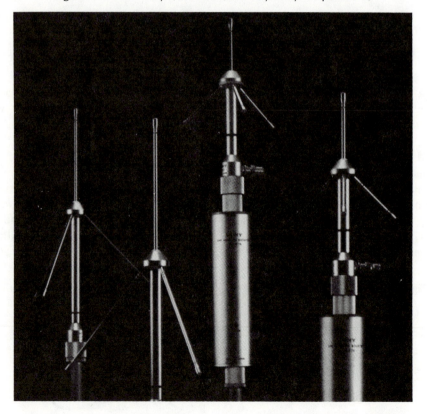

If mikes are permitted in the picture, the practice is to use the least obtrusive-sized mikes. If the mike is in close proximity to the performer, dynamics are used and may be hand held, desk-stand or floor-stand mounted, or worn, in the case of tie-tacks or lavaliers. If there are more than three performers whose mikes must be on, together, then cardioids are used to lower ambient sound and the chance of extraneous sound pickup, which is always present in large television studios.

If mikes are not permitted in the picture, then boom-mounted hypercardioid line mikes are used, since the mike may be 6 to 8 feet away from the performer. Two manually operated booms are often used, one covering the sound in progress and the other standing by in place for the next scene or point of action. The audio operator in the control room directs the positioning of the mikes by directing the actions of the boom operators in the studio. He quickly tells the boom operator if a boom mike has dipped down into the picture.

CARE OF MIKE CABLES

Mike cables should be neatly coiled and hung, away from foot traffic, when disconnected from mike and wall receptacle. Stepping on a connector, especially a male connector, may bend its shell out of round, and it then cannot mate with its receptacle. Abused cables or connectors invariably break down during times when easy replacement is least feasible. Make a quick visual check of all mikes and cables to be used when preparing to use a studio. Check to see that each mike to be used on a show is plugged into its respective receptacle. This avoids embarrassment when an unconnected mike fails to respond.

MICROPHONE STANDS

Mike stands or holders (Figure 11–62) are available in various configurations to be used as the need dictates.

Studio floor stands are of tubular concentric pipe attached to a heavy base plate at the bottom. The top is threaded with a standard $\frac{5}{8}$ inch–27 pipe thread, to which is attached one of many varieties of clamps or devices, which in turn holds the microphone. The clamps are often shock mounted to prevent mechanical vibration from being transmitted directly to the mike.

Portable floor stands, used for remote work, are lighter than their studio counterparts and have folding tripod legs instead of a heavy base plate.

Booms physically project the mike over and above the performer's head. There are two types. One, the baby boom, remains stationary and may be used by a single performer. The other requires an operator astride the boom and a second person to move it from place to place.

Desk stands, which employ a 3 to 6 inch threaded $\frac{5}{8}$ inch–27 tubular rod on a small base plate or tripod are used in all situations where the mike is placed on a table in front of the performer.

Mike clamps for remote jobs typically resemble a C clamp, usually with a gooseneck device attached, and are mounted by the operator to the sill of a lectern or any projection available, except another operator's clamp, to permit placing the mike in front of the user.

Shock mounts (Figure 11–63) are devices in which the microphone mechanically "floats" and thus is protected from mechanical handling noise.

Stage mount or mike mouse (Figure 11–64) is a foam brick round on top and flat on the bottom, developed by E-V, into which the mike is nestled, with its cable emerging like the tail of a mouse. It sits

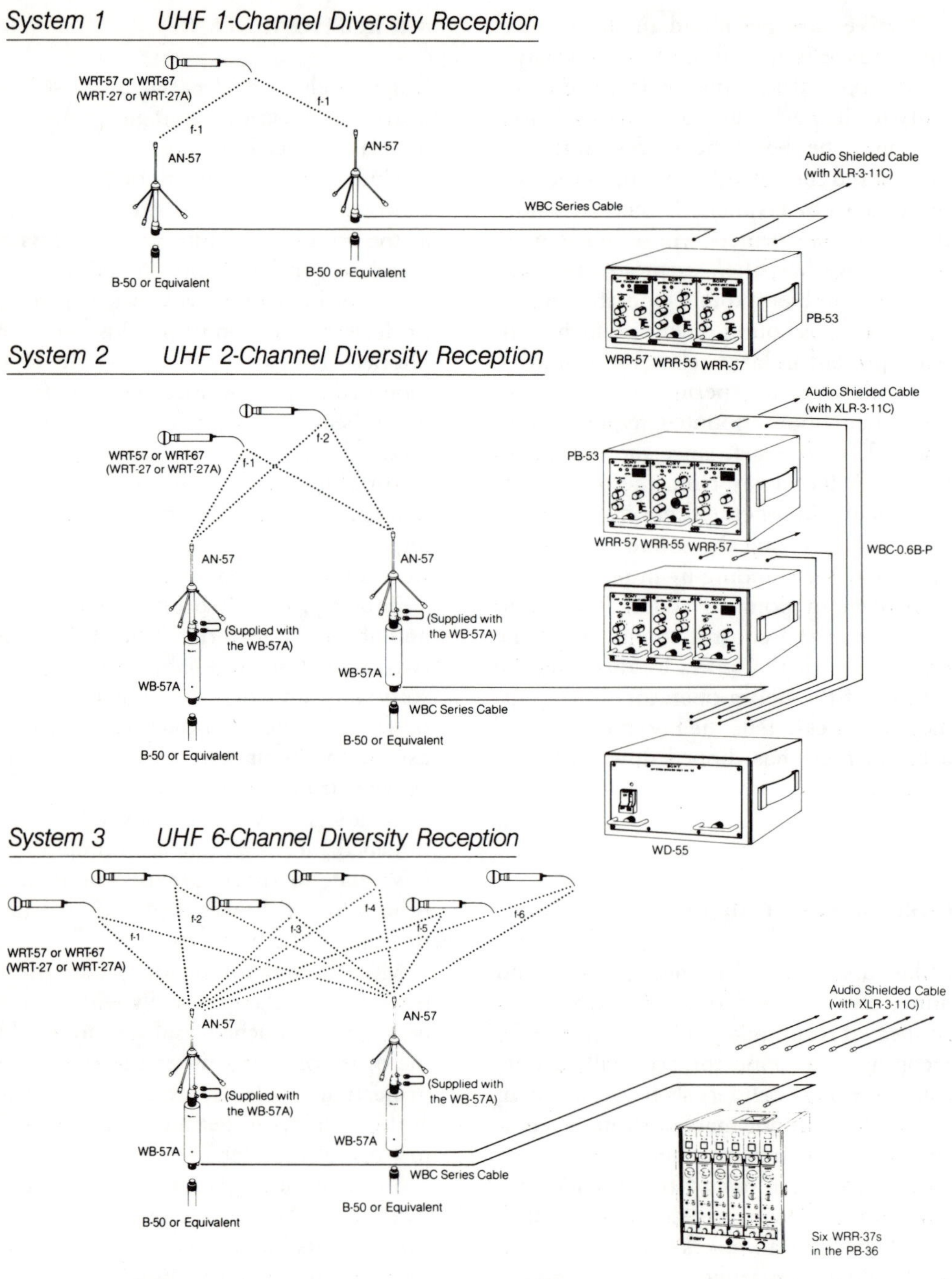

System 1   *UHF 1-Channel Diversity Reception*

System 2   *UHF 2-Channel Diversity Reception*

System 3   *UHF 6-Channel Diversity Reception*

Figure 11–61   Sony wireless mike systems 1 through 6. (Courtesy Sony Corporation.)

System 4   UHF Portable System

System 5   UHF Portable Diversity Reception

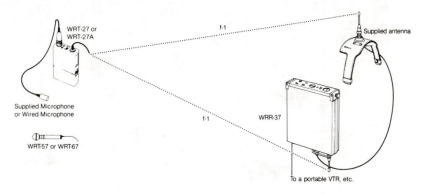

System 6   UHF 500mW Portable Transmitting System

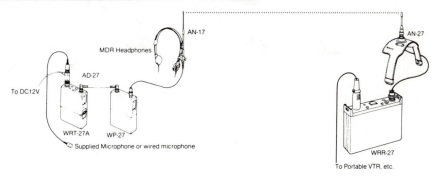

Figure 11–62   Mike stands and booms. (Courtesy Atlas/Soundolier.)

on the stage, picking up reflected sound directly, without the delay normally associated with reflected sound.

REVIEW QUESTIONS

1. What is a microphone polar pattern? Describe the four pattern types.

2. What is a response curve? What does it describe?
3. Explain popping and sibilance.
4. Describe the types of transducer element used in dynamic and in true condenser mikes.
5. What is the design flaw inherent in velocity mikes? For what type of voice are they best used?

Figure 11–63   Shock mount. (Courtesy Electro-Voice, Inc.)

Figure 11–64   Mike house stage mount, top *(left)* and bottom *(right)* views. (Courtesy Electro-Voice, Inc.)

6. Describe the three types of cardioid mikes. What is the principle employed in the Cardiline mike?

7. When do we use a lavalier? A transmitter mike?

8. Discuss on mike versus off mike. Define presence.

9. Describe the various types of microphone stands.

12

MICROPHONE SETUPS

This chapter discusses the arrangement of microphones for optimum sound pickup. It necessarily mentions specific situations and conditions, but since conditions and situations are subject to variation in real life, these setups must be considered starting points rather than rules.

First, let us look at mike setup complexity. Complexity is neither good nor bad but just is, and there are two aspects of mike setup complexity. The first aspect reminds the operator to keep it simple. Never use two mikes where one will do. The fewer mikes used, the less likelihood of opening the wrong one. Remember the mike proximity effect and opposite phasing with its cancellation effect, and so forth.

The second aspect has to do with what is necessary to get the optimum sound pickup. Sometimes it is necessary to use more than one mike to get the right pickup of even one musical instrument. Sometimes mikes must be angled more to avoid adjacent instrument pickup than to pick up the primary sound source.

Microphone arrangement, therefore, amounts to microphone compromise. We start to look at that compromise by briefly reiterating some of the points made in the previous chapter. To get the sound that we want, we must use a mike with the right directional characteristics and the right frequency response. The right pickup pattern is chosen as much to avoid un-wanted sound as to get the right sound. The right response curve is chosen to include or preclude bass proximity effect or high-frequency roll-off. The operator decides whether to use a directional or an omni mike solely in terms of sound avoidance. If it is not necessary to use a directional mike to avoid spurious sound pickup, then by all means use an omni. Omnidirectional mikes are better at avoiding mechanical noise, breath noise, and wind noise, and they have a wider, broader frequency response than directional mikes.

Once the proper microphone type has been selected, the operator must decide where to place it. The operator's choice of mike type is a difficult choice, despite what has been said above. This is because directional, mainly cardioid, mikes, are more expensive mikes, and one tends to want to use the most expensive tool available to do a job that will receive critical acclaim. The placement decision, which must be refined as many times as necessary, involves the distance between the mike and the sound source and, with a cardioid mike, the angle or cone of sound acceptance or avoidance.

Sound waves follow the inverse square law and decrease in intensity with the square of the distance from the sound source, so less sound is available for pickup as the mike moves away from the source. Too, the further away from the wanted sound source, the closer the mike gets to possible unwanted sources. Start the mike placement with the mike close to the source, remembering that some sounds have multiple sources (musical instruments), some sound sources are either very weak or very explosive, and some sources tend to move about (e.g., a jazz trumpet or a sax).

For the angle of placement of a directional mike, the operator must know the mike's characteristics from its polar pattern, the sound-radiating characteristics of the source (sharp like a drumhead or continuous like a cello), and the proximity and radiating characteristics of any unwanted sound that might be picked up.

Further considerations of distance and angle consist of whether the pickup is indoors or out—they have different ambience, and there is little or no reflection outdoors; whether the mikes may be seen; and whether the sound is that of a pop group or symphony orchestra, with the corresponding difference in sound blend.

An outdoor pickup with less reflected sound, even with a band shell covering all or part of the music group, will require tighter miking. A symphony orchestra performing indoors produces a blended sound that depends on the reflection of the hall. This requires miking with a pair of stereo mikes hung high above and about 15 rows of seats out in front of the orchestra. An unseen mike is used for a television play or movie, where display of a microphone would compromise the context of the piece.

In terms of music pickup, the mike's frequency response must always be wide enough to cover the range of the instrument being miked. Close miking has already been mentioned as essential in multitrack recording of pop groups to control the music from the recordist's point of view. Close miking is difficult with instruments that move about, so "contact" miking is often used. A miniature omni mike is either shock mounted on the instrument in close proximity to its sound source, or a contact mike is mounted on the instrument so that it directly picks up the instrument's vibrations. The output of this mike is often fed to a wireless transmitter, so when the musician moves about he is never off mike.

VOICE PICKUP

Single Voice Pickup

Our first consideration when we discuss voice pickup, is single voice pickup. We state this consideration in terms of the radio studio, for clarity, with adjustments to be made if the speaker is seen as well as heard.

Determine first whether the speaker desires, or needs, to sit or to stand. The nonprofessional speaker in a studio usually prefers to be seated to more easily handle script or notes.

Place the mike directly in front of the speaker, with its height adjusted at about chin level and the speaker talking either directly into, or at a 45 degree angle to, the transducer element as a starting point. Take a voice level and make final adjustment of the mike angle for optimum presence. Instruct the nonprofessional speaker to signal the control room visually if he desires to clear his throat or cough. Advise him to remove paper clips or staples from the script. Explain the noise made by rattling script pages or by bangles and bracelets. Ask the speaker to stay on mike and not to make extraneous noise such as foot tapping or table thumping. Finally, direct the speaker to watch for a starting hand signal, and advise him to give a visual signal when concluding.

Two-Voice Pickup

Exemplified by an interview, two-voice pickup is generally performed with two mikes, cardioids when necessary. The two speakers are arranged at a table, facing each other, a mike in front of each speaker, as in the one voice pickup. When taking levels, balance the voices at the console. If only one mike is used, position it so that it favors the weaker of the two voices. If a hand-held mike is employed, then the interviewer should hold the mike, remembering to angle it back and forth to pick up whichever voice is on, as the operator rides gain diligently between the two voices.

Groups of Voices

Round table news discussions and panel shows are examples of programs involving voice groups. With the participants in assigned seats, provide either a cardioid mike in front of, or an electret tie-tack mike on, each participant. Connect the mikes to the console so that they can be numbered and each assigned a pot from left to right within the group of participants, so that the pot corresponds to the participant. With careful emphasis on the numbering system, take levels around the group for total balance. During the performance, watch the group members carefully as the mikes are opened and closed. Leaving all mikes open makes the job easier for the operator but often raises the ambient noise level in the background of the program and may evoke comments from participants who think that their mikes are closed. Ride overall gain during the program with a submixer pot or the master pot.

MUSIC PICKUP

Monaural Music Pickup

For solo instrument pickup, a mike is placed at the optimum presence point for that particular instrument. This is determined, in a general sense, by the point on the instrument from which the direct sound emanates, that is, the bell of a trumpet, clarinet, or saxophone or the S hole of a violin or cello. Wind instruments need tighter miking than do brass. The grand piano, on the other hand, may be miked in several ways. Some operators mike a grand piano (with the lid opened at about a 45 degree angle) by placing a mike close to the instrument, perpendicular to the hammer line and aimed for the sound reflected from the lid. Other operators prefer to aim a dynamic mike (often a tie-tack electret) down into the third hole of the piano's metal bed or frame. A third method employs two dynamic mikes, one located at the second hole in the frame (from the pianist) aimed at the keyboard and at right angles to it, with the other mike located on the other side of the piano midway between the front and back of the lid, aimed at a 45 degree angle to the keyboard. The two mikes are fed to separate inputs to achieve a gain balance.

Miking too close to a violin, viola, or cello will pick up rosin squeal and other bowing noises; miking too close to woodwinds picks up reed and sputum noises. The placement of mikes must also consider the musician's arm movements, and, with a live audience, the mike should not hide the performer's face or instrument from audience view.

Stereo Music Pickup

For stereo pickup, the centrally located mike of the mono pickup becomes two mikes, one angled 45 degrees left and the other 45 degrees right, and each is fed to a separate input channel or a single stereo input channel. The side mikes normal to a mono pickup are either eliminated or used to augment the left and right central mike pickups.

Music Groups

In a monaural pickup of a music group, the prime concern is the pickup of the orchestral blend, as it would sound to a live audience. This is important and is mentioned because mike placement can color the sound blend as the blend is mixed in the console.

Miking a small music group of up to eight or ten instruments usually involves using three mikes, one front and center and one each front and side, all aimed at the orchestral group. The center mike is opened first (preferably in rehearsal), and then the left and right mikes are opened for music balance. Less gain will be required on the tympani side, and more gain will be needed for the reeds. The operator may consider moving the side mikes or angling them for optimum music balance. Solo instrumentalists are asked to step up to the center mike or to yet another mike that has been set in advance at the proper height for the instrument.

Miking a large orchestra monaurally is an expanded variation of the technique of small group pickup. One central mike is used for overall pickup, this time placed further in front than with the small group, or about 20 feet away on average, and 15 to 20 feet above the floor if possible. The mike may need to be hung, or it can be placed on a high stand if the audience will not find it objectionable. The left and right side mikes are often augmented by separate mikes on particular sections or instruments in the orchestra, such as the strings or woodwinds, due to the relatively low sound volume emanating from those instruments compared with the horns and tympani. Because of the complexity of the sound blend required for optimum pickup, this type of mike setup can spell chaos for the audio operator without adequate rehearsal time during which the mikes are moved and angled for correct music balance. After balance

is achieved, the operator controls overall gain with a submix pot or the master pot.

In this setup, vocalists, instrumental soloists, and announcements should be handled on a separate mike in front of or to one side of the orchestra. This mike should be opened only as needed and its gain adjusted separately as required to balance soloists against the orchestra.

The acoustic response of a mike pickup is affected by room parameters, the type of construction and angle of walls and ceiling, type of floor covering, and, most important, by the audience's absorptive effect. Keep this in mind when balancing mikes during rehearsal. The room will be much less reflective, that is, will be "deader," with an audience.

Multitrack Recording

In a multitrack recording pickup of music, the focus changes sharply from that of the mono music pickup. Here we wish to achieve a separate (unblended) pickup of each segment of the music group, each feeding a separate program channel and track, limited only by the number of program channels or tracks available. The tracks will be blended together later under controlled studio conditions. The operator must ensure that each segment of the group is picked up clearly, with as little overlap between segments as possible. This permits the later mixdown session(s) to be devoted to balance. A variation in the multitrack methodology embodies an instant mixdown to two tracks during performance. This complex system enables the producers and critical appraisers of the performance, such as the conductor and concert master of a symphonic group, to listen to a selection immediately after recording and determine whether it is a "take" or must be repeated by the orchestra.

At the opposite end of the musical spectrum, a rock group of four may be miked with as many as eight microphones if each performer in the group sings as well as plays. Each mike may feed to a separate track. Amplified instruments are picked up with a mike in front of the amplifier speaker. Contact mikes may be used on every instrument except the drums, and the vocal mikes may be hand held. Mike placement and the amount of mike cable employed depend on the movement of group members during the performance.

Choral groups are arranged by their directors so that they sing in discrete groups of altos, sopranos, tenors, and basses. Alternately, they may be arranged to sing in choirs of four within the choral group, with each choir of four having an alto, a bass, a soprano, and a tenor. If a chorus is arranged in separate voice sections, then each section can be miked separately, but choirs of four can only be picked up as a blend.

HOW MANY MIKES?

The only way to decide on the number of mikes is in terms of the minimum number. A stereo pickup must have at least two mikes. A set of drums should have a separate mike with separate response for every vibrating surface (drumhead, cymbal). Amplified instruments should have a mike at the amplifier speaker, usually a variable D cardioid aimed midway between the edge of the cone and the center. When a vocalist plays an acoustic guitar and sings, two mikes are indicated for the

same mike stand, one at mouth level and the other at the height of the hole in the guitar, pointed at the hole. The number of mikes for piano pickup has already been discussed.

SOUND AVOIDANCE

Before setting up mikes, especially indoors, stop everything for a moment and listen to the ambience, or room sound. Listen for sounds that usually are not heard separately by the ears because they are part of the background. Listen for fans (mechanical ones), air conditioning, creaking doors or floors, and reflection or echo. Clap your hands to hear echo. Remember that microphones, unlike ears, will not discriminate in favor of desired sound. Then choose and place the mikes to get the best compromise that good, patient audio practice will allow.

REVIEW QUESTIONS

1. Why do we try to keep it simple in monaural recording?
2. Describe and contrast the mike setups for one voice, two voices, and a panel.
3. Describe and contrast the pickup of a small orchestra monaurally and for a multitrack recording.
4. How does the mike balance problem differ between a monaural and a multitrack recording pickup?
5. Explain what is done during a mixdown session.

13

THE PROGRAM AS THE OPERATOR SEES IT

Now that the reader has some knowledge of the control equipment and its operation, we will integrate that operation into program formats as the audio operator sees them from his vantage point at the control board. We will observe the formats of a few fairly basic, but quite representative program types and indicate what the operator must consider and what his operations should consist of.

Note that this is not meant to be an exercise in programming, nor should these examples be considered to be solely radio or television formats. Our purpose in this chapter is to consider what the operator must think and do to use her equipment to maximum capability so that she may provide the programmer with the desired recorded or live material precisely when necessary.

The control room equipment available in each of our examples includes, in addition to the console, two reel-to-reel recorders, two cart playbacks, and two turntables.

THE NEWSCAST

Our newscast format will employ two mikes and their users, a news reporter,

and an announcer. The program format would look like this:

Ticker theme, tape cart

Live opening

First commercial, tape cart

Live introduction to news

Live news

Middle commercial, tape cart

Live news

Closing commercial, transcription

Live commercial tag

Live weather report

Live closing announcement

Ticker theme, tape cart

The operator meets with the program participants before the program. He is given the theme cart, the transcription—a large vinyl disk with many short cuts, and the commercial carts in the order that they are to be used, together with a copy of the format as indicated above, unless he is completely familiar with the program.

Now we follow the format through the program as the operator sees it, noting each audio operation performed.

Before airtime, place the theme cart on one playback and the first commercial spot on the other. Take levels on the announcer and newsreporter and on the theme and commercial inserts.

Ticker theme. At airtime, punch the start button on the theme cart and fade its pot up to full level.

Opening. After 5 to 10 seconds, fade the theme to background (BG) and open the announcer's mike key at the same time. The announcer reads the program opening while the theme is faded slowly out behind his voice. Stand by on the first commercial cart.

First commercial. At the end of the live opening, roll the commercial cart, pot open, and key it in. Clip the announcer's mike key.

Live news introduction. At the end of the cart spot, open the announcer mike key and clip the cart key. The announcer introduces the newsreporter.

Live news. At the end of the introduction, open the newsreporter's mike and clip the announcer's mike. Then remove the first commercial cart and replace it with the middle commercial cart. Note that the theme cart stays where it is, since no other carts are used on this machine and the theme repeats at the end of the program.

Middle commercial. At the end of the first news segment, start the middle commercial cart, key it in, and clip the newsreporter key. At the end of the middle commercial cart, open the newsreporter key and clip the cart key.

Live news. During this news segment, cue up the closing commercial transcription on a turntable.

Closing commercial. When the news is completed, roll the commercial transcription, key it in, and clip the newsreporter's mike. When the transcribed spot ends, open the newsreporter's mike and clip the turntable key.

Weather report. At the end of the weather report, open the announcer's mike and clip the newsreporter's mike.

Close. As the announcer closes the program, roll the theme cart and fade it in behind the announcer. On her last words, fade theme up full, clip announcer's mike, and hold theme at full level to program end. Fade theme out.

Note that in each instance cited in the newscast example, the upcoming operation is started before the ongoing one is completed. A mike or playback is always turned on before the previous one has been clipped. This standard practice is adhered to in all audio operation.

THE SPORTSCAST

A sportscast is essentially a news program with total emphasis on sports news and sports features. Our sportscast format includes themes; fanfares, which are used to engender excitement; and an actuality, a tape insert of an earlier recorded ballgame.

Here is our typical sportscast format:

Opening theme, record album, side 1, cut 3: football fight songs

Opening announcement, live

Theme

Live sports news

Actuality, tape

First commercial, tape cart

Tag, live

Fanfare, tape cart

Live sports features

Second commercial, tape cart

Closing announcement, live

Closing theme, record album, side 1, cut 5: football fight songs

As in the news program, the operator and talent meet before the show, and the operator is given a program rundown and all the recorded inserts in the order they will be used. Before airtime, levels are taken on the sportscaster and all recorded material.

In the control room, cue up side 1, cut number 3 of the record album. Then cue up the tape actuality insert and put the first commercial cart on one playback and the fanfare cart on the other.

Opening theme. As the program opens, fade in the theme to full level. On cue from the sportscaster, open his mike and fade theme to background behind him. At the end of the live opening, fade theme up full, clipping the mike at the same time.

Sports news. On signal, open the mike and fade theme to background. As the sportscaster begins to read the news, fade theme down and out behind him. Stand by on actuality tape insert.

Actuality. The sportscaster introduces the insert; on signal, roll tape and key in the insert. Clip the sportscaster's mike.

Commercial. At the conclusion of the actuality, punch on commercial cart, key it in, clip reel-to-reel tape key, and stop tape machine. Rewind tape insert and remove it from the machine.

Live tag. At the end of the cart spot, open the sportscaster's mike, clip cart key, and stand by on fanfare cart.

Fanfare. Sportscaster reads live tag to commercial. At the end of the tag, roll fanfare cart, fade it in to full level, and clip sportscaster's mike.

Sports features. At the end of the fanfare, open sportscaster's mike and clip cart key. Then remove the first commercial cart from its machine and cue up the second commercial cart on that machine and the closing theme, side 1, cut number 5 of the record album on its turntable while the sportscaster reads the sports features.

Second commercial. On cue from the sportscaster, roll the second commercial cart, key it in, and clip the mike key. The closing theme will be dead-potted. It runs 2 minutes, 20 seconds.

Closing announcement. At 2:20 from the program end, roll the disk, pot closed, key in program position. At conclusion of second commercial, open the mike key and clip the cart key. As the sportscaster reads the program closing announcement, on signal sneak the theme to background and allow it to build (fade it up) behind his voice. At the conclusion of the announcement, clip the mike and fade the theme up full. Theme will end at the precise time that the program is scheduled to conclude if it was dead-potted accurately.

THE DISK JOCKEY SHOW

The DJ program, from the operator's viewpoint, consists of alternately playing records (cuts on LP albums or 45s), commercial carts or cuts on transcription disks, and opening the mike for live lead-ins, introductions, and lead-outs.

The programmatic essence of a DJ show is "tight" audio control. The ins and outs of recorded music, and live or recorded speech, must tightly overlap. There may be no pauses, no holes, and no hesitations.

Here is our DJ program format:

Theme, cart
Opening announcement, live
Theme
Introduction, first record, live
First record, side 2, cut 3, in full
Lead-out first record, live
Commercial, cart
Tag, live
Commercial, cart
Tag, live

Commercial, cart

Introduction, second record, live

Second record, side 1, cut 1, sneak under

Commercial, cart

Commercial, transcription, side 1, cut 6

Introduction, third record, live

Third record, side A, 45 rpm, in full

Commercial, cart

Commercial, cart

Tag, live

Closing announcement, live

Theme, cart

As with other programs, the operator and the DJ meet before the program. The operator is given the record albums, single 45s, transcriptions, and carts, stacked in the order in which they are to be played, as well as a cue sheet or listing of recorded inserts in playing order by name, side number, cut number, and, often, timing.

In the control room, then, take the DJ's level and begin to set up the playbacks. This starts with the question, "What comes first, and next, and then?" Insert the theme cart on one playback, cue up the first record on one turntable, insert the first commercial on the second cart playback, and cue up the second record on the second turntable.

Theme. At airtime, hit the start button on the theme cart machine (key open, pot closed) and fade the theme in full. On the DJ's signal, open the mike and fade the theme to background.

Opening announcement. The DJ opens the show. On signal, clip the mike and fade theme up full. On the next signal, open the mike and fade theme down.

Introduction, first record. During the introduction of the first record, fade theme out under the DJ's voice.

First record. At the end of the introduction, release the record and pot it in so that the first note of music overlaps the DJ's last word. Clip the mike.

Lead-out record. During the first record, remove the theme cart and insert the second commercial cart on that machine. At the end of the record, open the mike and stand by on the first commercial cart.

First commercial cart. On cue, play the cart (pot open, keyed in) that starts with speech and clip the mike.

Live tag. At the end of the commercial, open the mike for the tag and clip the cart key. Stand by for next cart.

Second commercial cart. On cue, key in the cart (pot open) and clip the mike key.

Second record. At the end of the commercial, open the mike key and, on signal (key on, pot closed), sneak the second record (slow fade in) so that the vocal part of the record starts on the last word of the introduction. This is accomplished by timing the record from its first note to the start of the vocal and then watching the studio clock, talking with music under, until the start of the vocal. Clip mike. During the second record, remove the first commercial cart and replace it with the third cart, remove the first record from its turntable, and cue-up cut 6 on side 1 of the transcription.

Third commercial cart. On the last notes of the second record, punch on the third commercial cart (open pot, keyed in) and close the pot on the second record.

Transcription commercial. At the conclusion of the third commercial cart, segue to the transcription cut (pot open, keyed in) and clip the cart pot. Remove the cart and replace with the theme cart.

Live tag, closing-announcement. At the end of the transcription, open the mike key and clip the turntable key. The DJ reads the commercial tag and the show closing announcement.

Theme cart. Roll the theme cart, sneak it to background (key in, pot closed), and on cue bring theme up full to time.

Again, at the conclusion of the program, replace all records in their jackets and return all the recorded material to the DJ.

In this format example, we see that the operator literally has his hands full at all times. There is little time for relaxation during the program, and he often has scant seconds for cue-up, but he somehow always manages to be ready when the DJ signals for the next program insert.

THE PANEL PROGRAM

The panel program is, with the possible exception of a theme, an all-mike program. In our example there are four panelists and a moderator, each wearing a tie-tack mike, and a commercial announcer at a table mike.

Before the program, all participants including the operator meet in the studio. The panel participants and the moderator are seated in a semicircle, so that they can see each other, with the open end of the semicircle facing the control room so that the operator can see each one's face clearly. The operator places the mikes on the participants' tie, lapel, or dress so that they can turn to each other during the program without being off mike. The mikes are connected to their receptacles so that they are in the same numerical order as the panelists, from left to right. This enables the operator to rapidly identify mike with voice with pot as the program progresses. The announcer mike is connected numerically, following or preceding the participants' order.

Back in the control room, take levels, first on the announcer mike alone, then the tie-tacks. Take tie-tack levels, first with all five mikes open and pots adjusted for balance. Listen and watch the VU carefully at this point for ambient noise (room noise). If the level of ambient noise is low enough not to be disturbing, leave all five mikes open, occasionally balancing the pots as the discussion becomes heated. If the ambient noise level is high (this is indicated on the VU meter as a constant fluttering of the needle between -7 and -20 VU), perhaps due to air-conditioning rumble, then the alternate approach of two open mikes, panelist and moderator, is indicated. In either case, the moderator mike is kept open throughout the program. Watch the panel members intently, and as each prepares to speak or offer a rejoinder, quickly open the appropriate mike and close the mike of the panelist who spoke last. This method requires close observance of panelist facial expressions and fast reaction on the mike keys.

The format is as follows:

Opening theme, cart
Moderator, introduction to program
Commercial announcer
Moderator and panelists
Commercial announcer
Moderator and panelists
Commercial announcer
Moderator and panelists
Commercial announcer
Moderator
Closing theme, cart

Theme, cart. Before the program beginning, cue-up theme. The program opens with theme. Punch on theme cart, pot it in, fade up full, open the moderator's mike, and fade theme to background and out under moderator.
Moderator. The moderator sets the tone of the program, and, on cue, the operator opens the key to the commercial mike and clips the moderator key.

Commercial. The commercial announcer reads the sponsor's message. At the conclusion of the message, open the moderator mike key, clip the announcer mike key, and, if all panelists' mikes are to be kept open, stand by to throw all panelist mike keys on.

Moderator and panelists. On cue from the moderator, open either all panelists' mikes or the mike of the one to whom the moderator addresses the first question.

Commercial. Follow the miking procedure indicated until the moderator interrupts the discussion and introduces a commercial message. Open the commercial mike, and as the announcer reads the message, the panelists will invariably chat among themselves in (one hopes) hushed voices. The directionality afforded by a cardioid mike for the commercial announcer will help prevent pickup of this extraneous chatter.

Moderator and panelists. At the end of the message, open the moderator mike, clip the commercial mike, and stand by to open one or all of the panelists' mikes.

The above procedure repeats throughout the program until the moderator begins to close the show. On signal then, roll the theme cart (key open, pot closed) and slowly fade it in under the moderator. On the moderator's last words and cue, fade the theme up full to time.

THE TELEPHONE TALK PROGRAM

The telephone talk program can be considered a variation of the panel program, with the panelists in this case being members of the listening audience. They converse with the program moderator through their home telephones as they listen to the program. The home telephone in turn is fed to the console through a line input, so that the entire listening audience can hear both sides of the conversation. The aural quality of the incoming telephone line does not, of course, match the quality of the studio microphone, but it is accepted as necessary to the program format.

To mix a telephone line input with a microphone input from the studio, an additional piece of equipment, called a telephone interface (which includes a hybrid coil or, simply, a hybrid) interfaces between the telephone line and the console input. Figure 13–1 shows a telephone interface.

The SPH-3A is an interface that separates the "send" and "receive" audio components from a standard two-wire dial telephone circuit. The front panel controls are on-off phone switch, which connects or disconnects the SPH-3A to the phone line (not the power line on-off), unmutes the monitor speaker so the caller can be heard, and unmutes the caller audio going to the console input; the caller volume control, which adjusts volume of the incoming call; the monitor speaker volume control, which should be kept low enough to prevent the caller from sounding hollow or with echo as a result of acoustic coupling between the mike and the speaker; null, which adjusts the amount of rejection of the feed audio heard at the caller output, a one-time setup adjustment; and feed, another one-time adjustment of the amount of feed audio heard by the caller.

The rear panel has a monitor jack to a speaker that is used to listen to the caller; J6 is a remote control jack; line is for a telephone company modular plug to the telephone line; set is for a standard telephone; mix out is two outputs, one on XLR-3 and one on RCA phono jack, that contain a mixture of send and caller audio; caller out, on XLR-3, is the output of the caller's voice; and send is an XLR-3 input to the caller's phone line. The interface allows for compensation of the narrow bandwidth (200 Hz to 2000 Hz) of the telephone company circuit.

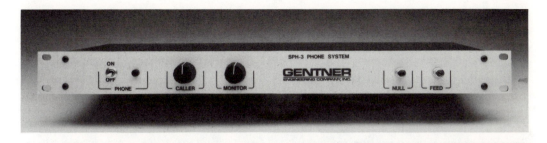

Figure 13–1   Hybrid coil—Gentner SPH-3A telephone interface, front *(top)* and back *(bottom)* views. (Courtesy Gentner Electronics Corporation.)

For optimum operation, the interface should either have or be used with a facility having "mix-minus." The SPH-3A includes mix-minus.

Mix-minus refers to what the telephone caller hears on his telephone, which is the program air mix minus his voice. To create such a mix, the studio sources are distributed to two busses: buss A to the air program channel and buss B to the send input of the telephone interface—the feed to the caller. The incoming audio return of the interface—the feed from the caller—is sent only to the A buss. The Henry MixMinus Plus (Figure 13–2) is a separate device that supplies mix-minus.

This arrangement prevents feedback problems that would occur if the telephone line were looped back on itself by having the return going back to the send. Further, the caller on a call-in program should always be advised to turn down the volume of his radio or television set for the same feedback reasons.

There is an inherent problem attached to call-in programming, which is that a

Figure 13–2   MixMinus Plus. (Courtesy Henry Engineering.)

telephonist might use profanity or other inappropriate language that, if aired, would contravene FCC rules and regulations.

To prevent inadvertently airing such comments, which would jeopardize the broadcaster's license, the incoming telephone line is fed first to a tape-delay device that permits the operator to screen and clip aurally if necessary the telephone comments before airing.

The device delays the conversation only momentarily (usually 7 seconds) and is apparent only to the telephonist and the control operator, not to the rest of the audience. The output of the delay device is then fed to a console line input. The operator controls the moderator mike and the telephone input and periodically inserts recorded commercials through his turntables and cart playbacks. During the program, he listens critically to the input signal to the delay device, prepared to clip its output instantly on hearing language that cannot be aired without penalty to the station.

The format is usually flexible and, for the most part, depends on the listener interest generated by the moderator and his caller participants. It might look like this:

Program opening, moderator, live

Commercial, cart

Moderator and caller, live

Commercial, cart

Commercial, transcription

Moderator and caller, live

Moderator. At airtime, open the moderator mike. The moderator begins the program by offering a controversial comment and then asks for listeners to call, giving the telephone number assigned to the program.

Commercial, cart. On cue roll commercial cart and key it in. At the conclusion of the cart, open the moderator mike and clip the cart key.

Moderator and caller. As the moderator accepts the first phone call, key in the telephone input. Adjust phone input levels and listen to the incoming conversation. The moderator may take several calls, keeping the program lively, and then indicate a standby for the next commercials.

Commercial, cart. On cue, roll the next commercial cart, key it in, and clip the phone input key and moderator key.

Commercial, transcription. At the cart conclusion, roll the transcription, key it in, and clip the cart key. When the transcribed commercial ends, open the moderator mike, and clip the turntable key. The moderator then accepts additional calls, the line input key is opened, and the program continues in this vein, often for hours at some stations, following the same pattern.

THE DRAMATIC OR DOCUMENTARY PROGRAM

The documentary is now rare in radio. This program format requires that the operator be able to follow and use a script.

Some time in advance of the program, which is often recorded rather than presented live, the operator, a director, the cast, and commercial announcer meet in the studio. The operator is given the program script, the recorded themes, music bridges, mood music, curtain music, and any recorded sound effects, all in the order in which they will be used, together with a rundown sheet listing the recorded inserts in the same order.

The operator scans the script to determine what equipment will be needed on the program. If reverb, echo, or other effects equipment is required, then that device may have to be patched into the console.

During a program rehearsal, microphones are set up, two or more as required for the cast and a separate one for the commercial announcer. Levels are taken on the participants, and the level settings are marked in the script. Of particular importance are abrupt level changes, such as hushed whispers and shouts, which are carefully marked in the script just before the places where they occur. This enables the operator to fade the indicated mikes quickly up or down as the occasion demands during the program. As a part of the rehearsal, all the recorded material is inserted, in its proper sequence, to give the operator a "feel" for the requisite timing for each insert. At this time, he marks the insert level settings in his script as well.

The format for our dramatic presentation is as follows:

Theme

Opening Announcement

Theme

Opening commercial

Lead in to narrator

Narrator

Mood music

Cast, first act

Bridge music

Middle commercial

Bridge music

Cast, second act

Curtain music

Narrator

Closing commercial

Theme

Closing announcement

Theme

Previous to airtime (or recording time), cue-up the theme and mood music inserts.

Theme. Roll the theme, pot it in full to establish, and open the announcer mike, fading theme to background.

Opening announcement. Hold theme under while announcer introduces the program.

Theme. At the conclusion of the introduction fade theme up full and clip announcer mike. At theme end, open the announcer mike and close the theme pot.

Opening commercial. The announcer reads the opening commercial and introduces the narrator. Open the narrator mike and clip the announcer mike.

Narrator. The narrator sets the scene of the dramatization.

Mood music. During the narration, at the place indicated in the script, sneak the mood music and hold under. At the end of the narration, fade the music up full, open the cast mikes, and fade the music out under the cast.

Cast, first act. As the cast begins the dramatization, remove the theme and mood music from their playbacks and cue-up any sound effects or, if there are none, the mid-program bridge and the curtain music. If playbacks are available, cue-up the inserts for the second act at this time.

Bridge music. As the scenes in the dramatization change, play the bridge or transitional music as indicated in the script to effect the scene changes.

Curtain music. As the first act is ending, sneak the curtain music in under the final words and up to full level.

Middle commercial. When the curtain music concludes, open the announcer mike and close the curtain music pot. The announcer reads the commercial copy.

Curtain music. At its conclusion, roll the curtain music for the second act of the dramatization, fade it up full, open the cast mikes, and fade the music to background and out under the cast.

Cast, second act. The cast performs the second act of the dramatization. Remove the bridge and curtain music inserts and cue-up sound effects, bridges, and curtain music for the second act as well as the closing theme.

Curtain music. As the second act is concluding, sneak the curtain music, fade up in full, clip the cast mikes, and fade out the curtain music under the narrator who wraps up the dramatization.

Narrator. When the narrator ends, open the commercial mike and clip the narrator's mike.

Closing commercial. At the end of the commercial, roll the theme, bring up full, then fade down under the announcer, who closes the program.

Theme, closing announcement. As the announcer concludes, fade theme up full, clip announcer mike, and play theme full to time.

The director was mentioned at the outset of this program example. It is he, of course, who cues both the cast and the operator, although the experienced operator often anticipates the cues and thus keeps the production aurally "tight."

REVIEW QUESTIONS

1. From an audio operator's viewpoint discuss in length the admonition, "Watch the mike, watch the script, watch the clock, keep it tight."

2. Why should the operator be present during rehearsal of a music program or dramatic presentation?

14

STUDIO–
CONTROL ROOM
COMMUNICATION

The need for communication facilities between the studio and the control room became apparent during our discussion of control consoles. The program participants on either side of the double glass window separating a studio and its control room can communicate with each other using the control console's audition and talkback systems or a separate production intercom, provided there is no live mike in the studio.

PRODUCTION INTERCOM

In most cases, the production intercom can be thought of as having three parts. All systems have a "master station," which is part one. The master station includes the system power supplies; the configuration switches, which determine who may talk with whom; and other auxiliary controls common to the entire system. All systems have one or more local stations. These local stations are located wherever an operator, or other program personnel, will be during a production. Usually, in a television studio, there is one station for each camera and one for the director, the

switcher, the audio operator, boom operators, tape room, floor director, stage manager, light board operator, and sometimes the producer.

Part two is the local station. It usually has connectors available to "daisy chain" several units, a unit being a headset with earphone and mike combination. The local station has a local volume control and some form of "attention" signal—either a flashing light or a tone in the earphone. Local stations may take the form of belt packs, wall-mount speaker-headset stations, paging stations, or rack-mounted stations.

The third part, which may be optional, is the interfaces that connect differing systems to each other. Interfaces may be used to integrate wireless intercom systems, telephone lines, and the master station.

HAND SIGNALS

How is the communication thread extended to the studio when there is live mike(s)?

A system of hand signals has been developed for radio broadcasting, where there is usually visual contact through the studio–control room window. This hand-signal system was extended for television, where the control room is kept semi-darkened, by the production intercom through which the program's director gives cues to the camerapeople and to a floor manager or floor director in the studio. The floor manager in turn relays those cues to the talent as he stands next to the camera giving hand signals. Any boom operators in the studio are cued by the console operator through their headsets.

The hand signals described and illustrated on the following pages are not universally employed by every station. Indeed, many stations have developed their own signals to cover particular program situations. The signals presented here are representative of those used at many stations in the United States.*

Note that the photographs are precisely that—pictures, and as such they cannot depict movement of the hand or arm. If the hand in the picture is not exactly where the text says it should be, it is in motion and will get there.

We have grouped the pictures in categories roughly related to the signal use.

Figure 14–1 depicts "attention." One hand is raised above and over the head, palm extended forward and waved from side to side to attract attention.

Figure 14–2 depicts "stand by." One hand is raised over the head, palm forward, and held stationary. This is a preliminary signal and indicates that another signal is about to be given.

*A curious aside is that some signals in common use in the United States, such as the familiar OK, made by touching the tips of the forefinger and thumb together with the three other fingers extended upward and hand held above the head, or the equally familiar you're on signal made by pointing the forefinger, are considered severe insults in other parts of the world. The OK signal should never be given to a Brazilian, nor the single fingered "You're On," to an Asian.

Figure 14–3 depicts "watch me." It is made by pointing to one eye with the index finger.

Figure 14–1   Attention.

Figure 14–2   Stand by.

Figure 14–3   Watch me.

Figure 14–4 says "you're on." This is perhaps the most familiar signal in broadcasting. It is made by pointing at the talent with either the index finger or with the open hand, palm up. It is always preceded by the stand by signal and may be given to talent at the mike or to the operator, signaling him to perform some predetermined operation.

Figure 14–5 depicts "audition my mike." The talent at the mike forms the letter A with the thumbs and index fingers of both hands to indicate to the operator that he wishes to talk to the control room but not on air. The signal is not used when there is a live mike in the studio.

Figure 14–6 illustrates "open my mike." This says put this mike on the air. It is given by the talent at the mike by rapidly pointing at the mike several times.

Figure 14–7 depicts "clip my mike." It is made by drawing either the index finger or the open hand, palm down, across the throat. This signal should be preceded by the stand by signal but is often given as an emergency signal (e.g., a cough coming on).

Figure 14–4 You're on.

Figure 14–6 Open my mike.

Figure 14–5 Audition my mike.

Figure 14–7 Clip my mike.

Figure 14–8 depicts "cut." This is identical to the clip my mike signal. It orders immediate termination of whatever is in progress.

Figure 14–9 means "move closer to the mike." It is made with both hands raised to chin or eye level, palms facing each other and rapidly, repeatedly moving toward each other without touching.

Figure 14–10 means "move away from the mike." Both hands are at eye or chin level, palms facing outward or away from each other and hands moving rapidly and repeatedly away from each other.

Figure 14–11 illustrates "speed up reading pace." One hand is raised to eye

Figure 14–10 Move away from the mike.

Figure 14–8 Cut.

Figure 14–11 Speed up (reading pace).

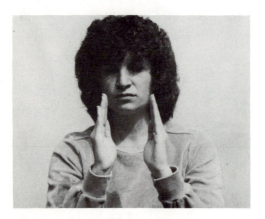

Figure 14–9 Move closer to the mike.

Figure 14–12 Stretch.

level, the index finger points to the talent, and then the wrist bends and rotates, with the index finger still pointing at the talent. The wrist rotation is rapid and in a large circle for a large increase in reading pace, or it is slow and in a small circle for a small reading increment.

Figure 14–12 means "stretch." This means decrease reading pace, or slow it down. Both this signal and the speed-up signal are used to control the pace and thus the timing of the program. Stretch is made with both hands at the chin, at eye level or above the head. The hands, palms in, are drawn apart as if there were a rubber band between them—wide apart for a large stretch and perhaps an inch or two apart for a small decrease in reading pace.

Figure 14–13 shows "segue." One hand is raised to eye level, palm in, with the index and second fingers intertwined and pointing upward.

Figure 14–14 depicts "crossfade." Both hands are raised to eye level, palms down, and then are lowered across the body on an arc so that they cross each other at chest level.

Figure 14–15 means "play the insert" whether disk, tape, or cart. Since it is given only to the operator, it should not be confused with the speed-up signal, which is similar. The hand is at eye level in a fist, with the wrist bent down and the index

finger slowly describing a circle. The hand then indicates the stand by, followed by the you're on signal.

Figure 14–16 shows "fade-in." This tells the operator to bring in music (or background sound) or to fade it up. It is given with one hand palm up, which is brought up from the side either directly or in a slow spiral until the sound level (on the monitor) is at the desired level. The hand then stops rising.

Figure 14–17 depicts "sustain level." Preceded by the fade-in signal, it is given to keep the level where it is. It is made by describing a slow circle with hand, palm up, without raising or lowering the hand.

Figure 14–14   Crossfade.

Figure 14–13   Segue.

Figure 14–15   Play the insert.

Figure 14–16   Fade-in.

Figure 14–18   Fade-out.

Figure 14–17   Sustain level.

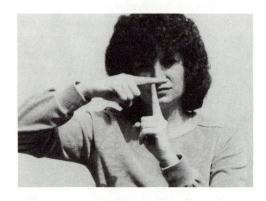

Figure 14–19   Play theme.

Figure 14–18 means "fade-out." From the sustain level hand position, the hand turns over, palm down, and the hand lowers directly or in a spiral to the side. If the music is to be taken out completely, then the hand snaps to shoulder level in a fist with the thumb pointing backward.

Figure 14–19 depicts "play theme." With both hands at the chin, at eye level, or above the head, the index fingers form the letter T. When the signal is acknowledged, one hand moves to "stand by" and then to "you're on."

Figure 14–20 means "thirty seconds." A time signal, made similarly to the theme

Figure 14–20   Thirty seconds.

signal, but this time the index fingers form a + sign rather than a T. Do it smartly so it will not be confused with the theme.

Figure 14–21 depicts "fifteen seconds." One hand is raised in a clenched fist, fingers forward, at eye or over head level. This signal was adapted from a musician's signal "play the sock chorus."

Figure 14–22 means "station break." Both hands at chin level or above the head make the motion of breaking a twig between clenched fists.

Figure 14–23 means "on the nose." This indicates that the program is right on predetermined time schedule or that talent has just performed a task well. It is made by rapidly tapping the nose with the index finger.

Figure 14–24 says "read the heads," or "take it from the top." This is an order to the talent to read the headlines or to restart, whichever is appropriate. Made by tapping the index finger repeatedly on the forehead or on top of the head.

Figure 14–25 says "read the opt" (optional item). This is made by repeatedly pointing the index finger at eye level to a real or imaginary script.

Figure 14–26 says "read the short piece." This is given to the talent to in-

Figure 14–21   Fifteen seconds.

Figure 14–23   On the nose.

Figure 14–22   Station break.

Figure 14–24   Read the heads, or take it from the top.

Figure 14–25   Read the opt.

Figure 14–26   Read the short piece.

dicate that she should read the shorter of two optional items. It is made with one hand at eye level forming the letter C or a bracket.

Figure 14–27 means "OK." The signal for OK is made by thumbs-up. The hand is held in a fist at eye level with the thumb upward.

Figure 14–28 says "wrap it up." It is given to close the program and made by raising both hands to eye or chin level, palms downward, and rotating the hands about each other.

CUE SHEET

Another important element of communication in radio broadcasting is the cue sheet, particularly on programs where only two participants are involved. When there is only the operator in the control room and a DJ or newsperson in the studio, a cue sheet is used. It is a run-down on paper of the recorded inserts employed on the program in the order in which they will be played that specifies in each case whether disk, tape, or cart; side number, and cut number; running time; and, for taped speech inserts, an in cue and out cue. Figure 14–29 shows a cue sheet.

A cue sheet should be prepared for the

Figure 14–27   OK.

Figure 14–28   Wrap it up.

✓	Insert #	T D C	Side #	Cut #	S or M	Time	Bring In F,F/U,S, BG,DP	Artist or In Cue	Title or Out Cue

Station WXZY

Date_____

Program_____

Director_____

Audio Operator_____

Legend: T-D-C- Tape, Disc or Cart
S or M Begins with Speech or Music
Bring in: Full, Full then Under, Sneak
          Background, Dead Pot

Figure 14–29   Cue sheet.

operator by the writer of the program if that program has more than one or two recorded inserts. The cue sheet permits the operator to prepare inserts for the program at his own pace, limited only by the number and type of playback equipment available to him. The cue sheet described here was designed to incorporate the information needed most by the operator, with a minimum of preparation time required of the person preparing it.

The first column has as its heading a check mark, which is used by the operator, after the fact, to indicate that the insert has been used. The second column list the inserts in numerical order of use.

The next column is labeled T D C. Insertion of the indicative letter tells the operator that the insert is tape, disk, or cart. The next two columns indicate side number and cut number, if either is applicable. S or M in the next column tells the operator whether the insert starts with speech or music, and the column following indicates the running time of the insert in minutes or seconds. To cue the operator as to how the insert should be brought into the program, the appropriate letter (F, full; F/U full, then under; S, sneaked; BG, background; DP, deadpot) is entered in the next column. The last two columns, which are larger than the others, are used to indicate either the artist and title of the insert or the in cue and out cue, whichever is applicable.

When an insert is to be eliminated from the program for any reason, but perhaps still appears as a cut in the middle of a large reel of tape inserts, then a line drawn through that insert on the cue sheet tells the operator to skip that insert. This feature also saves time for production people in prestudio program preparation.

REVIEW QUESTIONS

1. Why are hand signals used in broadcasting?

2. Who gives hand signals, and to whom are they given?

3. Explain the difference in use of the clip mike and the cut signals.

4. Why do we use different signals for record and for transcription?

5. Describe the difference between the signals for theme and for 30 seconds.

6. Why are hand signals not given directly from a television control room?

7. How are hand signals used in a television studio?

8. Describe a cue sheet and its use.

15

REMOTE BROADCASTS AND RECORDINGS

Remote broadcasts and remote recordings are, for the most part, coverage of special events or of actualities by a station that intends to broadcast those events or actualities either as they occur in real time or on a tape-delayed basis in full or in part.

An archaic term for remotes is NEMOs (not emanating main origination), since this was the way that they were entered in the station's log. British operators call them OBs (outside broadcasts).

Remotes provide an additional challenge to the operator's skills, since they are performed "in the field" with portable equipment and little chance for retakes. The equipment employed for remotes should therefore be checked very carefully for operational capability before it leaves the station, since in the field there will be no opportunity to patch around a faulty component. Also, equipment setups and arrangements on location take considerably more time than equivalent setups in a studio. A good rule of thumb to follow is to figure the amount of time necessary to travel to the remote site, to make arrangements and set up equipment, and then double it. All sorts of pitfalls may arise, from heavy traffic en route, to the lack of AC power on site, to a delay in telephone company (Telco) line installa-

tion, to a misunderstanding of the permission from the participants to allow recording or broadcast, to interunion misunderstandings, to an infinite number of other delay factors, all of them solvable given sufficient time in advance to make arrangements on site.

The operator on remotes will work outdoors in the hottest, the coldest, and the wettest weather. His patience will be tried in myriad ways. He will, as well, come in contact, often under very enjoyable circumstances, with some of the most important people in his community and in his nation.

The equipment used on field jobs or remotes all focuses around a small portable, limited function mixer or a portable tape recorder. Each of these devices must have at least one, and preferably two, 600 ohm line outputs capable of feeding program to a terminated telephone company line and must also have an input capability sufficient to handle the number of mikes necessary to mike a special event feed properly. There should be, additionally, a high-level input available to handle inputs from other mixers or to handle a very hot feed from a PA system.

Since most portable hand-carried tape recorders have limited input capability—generally not more than two inputs—a mixer or mixers are often used to feed additional needed inputs to the recorder.

The rest of the equipment needed to do a field job is mentioned next, and its use is described as the chapter progresses.

EQUIPMENT NEEDED

The following items are needed in addition to a mixer:

1. Microphones: a sufficient number of the particular type(s) for the job. If only one mike is needed, take two along. Alternately be prayerful that a colleague from another station is covering the same event, and that he will give you a feed from his equipment. Always be ready to return that favor.
2. Mike stands: a sufficient number of portable floor stands, table stands, podium clamps, tie-tacks, and mike-hanging devices to do the job.
3. Mike cables: mike extension cables, AC power cables, and a telephone company cable pair or a wireless mike system with fresh batteries.
4. Earphones: a high-quality pair that do not irritate the ears with extended wear or high-quality loudspeaker(s) for music jobs. The speakers should have handles on two sides and casters mounted on the treble end so they can be set upright with "wheels up" and the bass end down.
5. Connectors-adapters: a set, usually made up as needed by the operator, but also available from Switchcraft— XLR-3 male to male, XLR-3 female to female, XLR-3 to phono plug, phono plug to miniphono plug, and many others that allow the operator in the field to connect her equipment to or from just about any one's as needed.
6. An interface coupler for connection to telephone company lines.
7. A flashlight and a set of tools that should include, at a minimum, large and small screwdrivers (flat and Phillips head), diagonal cutters, long-nose pliers, gas pliers, and soldering equipment.
8. A roll of 3-inch wide "gaffer," or duct tape, which is used for temporary insulation of temporary wire connections and to tape down cables running across a foot traffic area on a stage, a ballroom, or a gymnasium—wherever cables could create a danger to people and equipment. The tape should be nonreflective and gray or black. If you run out of tape on the job, ask a colleague from another station, rather than risk endangering others. Gaffers tape, masking tape, or plastic cable clamps are also used after a remote to secure mike cables that have been carefully coiled and wrapped.
9. A suitcase or other conveyance large enough to carry everything but the floor stands and loudspeakers, and a two-wheeled cart or dolly to move everything from automobile to jobsite.

Remote Mixer

Truly portable mixers rarely have more than a five-input capability, but they usually can be "ganged," "stacked," or connected together to double or triple the number of available inputs. They are designed without input keys so that all operations in use are potting operations. One or more of a mixer's inputs is usually switchable to either mike level and impedance or to line level and impedance. There are no cuing or talkback facilities, and monitoring is provided by a headphone jack across the output of the line amplifier.

The input and output connections to the

mixer are made on site by the operator, generally to XLR-3 receptacles and binding post connectors located on the back of the mixer. The unit often is powered alternately by either AC power or self-contained batteries, with a provision that if the line power fails, then the battery power will be automatically switched in without loss of program.

The operating face of the mixer will have a pot for each input, a VU meter, and a master pot. There will be an earphone jack located either on the front or back of the unit. This jack may be used to feed program to earphones or to an amplifier-driven loudspeaker on music assignments. The frequency response of earphones is limited by the physical size of their transducing elements. Earphones are particularly bad in the low-frequency audio range, which, in turn, limits the operator's ability to hear the full range of an orchestra. This limits his ability to achieve full orchestral balance with his mike setup.

When a loudspeaker is used, the speaker and, indeed, all the mixing equipment must be located far enough away from the stage area, perhaps in a nearby soundproofed room, so that it may be operated at fairly loud volume without its being heard by either the musicians or the audience. The operator, on the other hand, must hear only the loudspeaker and not the direct sound of the orchestra. This situation, if optimized, prevents the operator from seeing the orchestra and visually judging mike-to-performer relationships. That condition can be eased somewhat by placing a small video camera in front of the orchestra, feeding a video monitor at the audio operator's location.

Figure 15–1 is a basic block diagram of such a mixer. Note how similar it is to the control consoles that we looked at earlier. Also note how the inputs are switchable in level and input impedance and how the high level input bypasses the preamp and feeds through the pot to the mixer buss.

We will now examine some of the refinements to that basic mixer found in professional versions.

Shure M267

The Shure M267 (Figure 15–2) is perhaps the most popular remote mixer used in broadcasting. It weighs 5 lb 2 oz and has a response of 30 Hz to 20 KHz.

The front face of the M267 has four rotary input pots, a master pot, and the VU meter. Above the input pots are low-cut filter switches for each input. The filters provide low-frequency roll-off to reduce wind noise or other low end noise. Above the master pot is a limiter in-out switch that turns on a fast-acting, peak-responding limiter circuit that cuts overload distortion during loud program intervals without affecting normal program levels. The peak LED indicator to the right of the limiter switch shows limiter operation with the limiter in and operates (flashes) when program levels approach overload with the limiter out. The LED is much faster than the VU meter and is activated by transient audio peaks.

A 1000 Hz tone oscillator is activated by a switch between and below pots one and two. Its level is controlled by pot one, and its signal appears on the line and mike inputs and on headphone and mix buss connectors. The oscillator should therefore be turned off when not in use.

The M267 power on-off switch is located to the upper right of the VU meter. In addition to AC power, the mixer can be operated from an internal battery supply consisting of three 9 V alkaline batteries. Access to the battery compartment is on the bottom of the mixer. With batteries installed in the compartment, the M267 will automatically and silently switch over to battery operation should the AC fall below a suitable level or fail completely. If AC power fails, the operator is notified as the VU meter illumina-

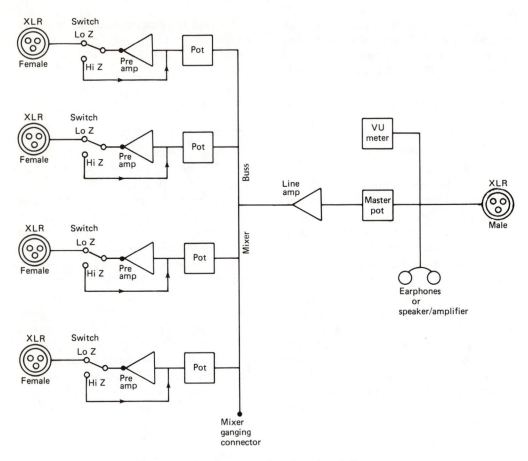

Figure 15–1   Remote mixer functional diagram.

tion lamps go out. Battery condition can be determined on the VU meter by using the battery check momentary switch located to the right and below the master pot. To use the battery check switch, the power cord is disconnected, the power switch turned on, and the BATT CHECK switch is depressed. A new set of batteries will show about a +2 indication on the VU meter. Battery condition is good if the reading is above 0 VU on the meter.

Battery life is about 20 hours of operation at +4 output, continuous use. Simplex-powered mikes or high-level headphone monitoring will increase the drain on the battery supply.

The rear face of the M267 mixer has an output XLR-3 connector, followed by four input XLR-3s. Each connector has above it a line-mike switch, which determines whether that input or output shall be at line level and impedance or mike level and impedance. Each input connector is directly behind the pot that controls the input. To the right of input 1 is the mix buss pin-jack connector, which permits stacking or ganging mixers for additional input capability. With two M267s stacked or multed through this direct access to the mixing busses, the operator has two independent master gain controls, two isolated line amps, and eight individually

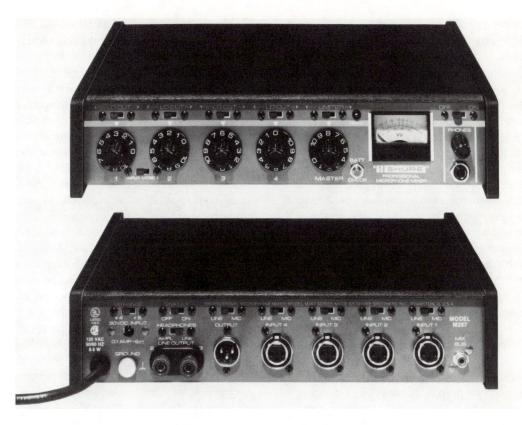

Figure 15—2   The Shure M267 mixer, front and back views. (Courtesy Shure Brothers, Inc.)

controlled inputs. Since the busses are directly paralleled, when connected, a 6 dB gain loss will occur, and the master or input controls must be increased to compensate for that loss.

To the left of the XLR-3 output connector is a binding post connector pair for line output, which is in parallel with the XLR-3 and can be used simultaneously. The binding posts are marked 2 and 3 to correspond to pins 2 and 3 of the XLR-3 and are used in conjunction with the ground post to their left. Incidentally, when the M267 is used with AC power, its power supply is energized when the AC cord is connected, and the unit should be grounded (earthed) for the operator's safety.

Above the line-output binding posts is the headphone amp-line switch, which permits choosing between the headphone amplifier (AMPL) and the line for operation when talkback monitoring is needed on remote broadcasts. The headphone output appears on the front panel to the right of the VU meter, below the headphone level control. The headphone output level is high enough in the AMPL mode to use as an additional but unbalanced line output.

Above the headphone switch on the rear panel is a simplex (phantom power) switch to provide power to condenser mikes if any are used with the M267. Simplex power is not applied if the input mike-line switch is in the line position.

To the left of the simplex switch is a VU range switch that selects either a +4 or +8 dBm output at 0 VU meter indication. This switch changes the meter indication but does not change the actual output level. The meter is calibrated for use with a 600 ohm terminated line. The +4 range is recommended for normal use, to provide approximately 14 dB of headroom from operating level to clipping level. The +8 range should be used with the limiter in operation, because limiting action begins at +7 VU. Below the VU range switch is a 30 V externally fused power plug input for powering the M267 with low external DC voltage.

Shure FP31

The Shure FP31 mixer (Figure 15–3) is a three-input mixer that was designed for ENG use. Its front view is shown in its protective pouch. Three XLR-3 connector inputs on the left side and two XLR-3 outputs on the right side are each switchable (above the connector) for mike- or line-level operation. A master pot is found to the right of the input pots. Built in to the FP31 is a "slate" mike for emergency field use. The mike is controlled by a slate pushbutton above and between input pots two and three, which also activates a one-second timed low-frequency slate tone. There are switchable low-cut filters above each input pot. A tone oscillator is activated by a pull switch incorporated in the number 1 input pot. There is a switchable limiter, with peak overload LED indicator. The two headphone outputs can be used as additional line feeds, and there is switchable phantom powering for the inputs. The FP31 is powered by three 9 V internal batteries, which can be tested without program interruption. Supplied with the FP31 is a removable shoulder strap and a carrying pouch that permits easy access to every mixer function; it al-

Figure 15–3 The Shure FP31 mixer, front and side views. (Courtesy Shure Brothers, Inc.)

lows for Velcro-connected piggyback operation of the mixer on a tape recorder. A similar mixer, the FP32, is available for stereo operation.

Portable Tape Recorder

Nagra Recorders

Nagra tape recorders (Figure 15–4) are literally hand-made in Lausanne, Switzerland, by the Nagra Kudelski Company. They are unqualifiedly the very best portable recorder-reproducers made. They are expensive (in the $8000 + range for the model 4.2), and they are extremely durable, trouble-free machines. "Nagra," in rough translation, means tape recorder in Polish, which is the mother tongue of its designer-manufacturer and emigre to Switzerland, Stefan Kudelski.

The Nagra 4.2 can be operated on a self-contained battery supply or on AC power. To operate on AC requires an external accessory, the ATU, which plugs into a "Tuchel" DIN type European connector on the right side of the Nagra. Twelve D cells are in the battery supply (alkaline type recommended), which is located in a compartment in the bottom of the recorder. To change batteries, release the compartment-cover-holding screw with a coin and remove the cover. Observe the correct battery polarity indicated in the compartment when replacing batteries.

The top panel of the Nagra 4.2 has its feed reel on the left and take-up reel on the right, with a speed-equalization switch

Figure 15–4   The Nagra 4.2 recorder-reproducer. (Courtesy Nagra Magnetic Recorders, Inc.)

in between the reels. There are separate switch settings for standard and low noise tape at each of the three tape speeds of 3.75, 7.5, and 15 ips. The tape path of the 4.2 is controlled by a control lever at the lower right of the top panel. Operating the lever toward the back of the recorder only partially closes the tape path and partially engages the capstan and pinch roller. These functions are completed when the main mode function switch directly below the lever on the front panel is turned to either the record or play modes.

To load tape, the control lever is pulled forward to the front of the machine, which opens the tape patch.

The tape path is not completely closed unless the recorder-reproducer is in an operating mode, to protect the pinch roller from being flattened in one spot from constant contact with the capstan.

To rewind tape, the control level is pulled forward as when loading tape, with the mode function switch in the playback or loudspeaker position and the fast-forward, fast-rewind toggle switch on the left front corner of the top panel in the left or rewind position.

To fast-forward the tape, throw the toggle switch to the right, with the mode function switch in the loudspeaker position. The 4.2 will not operate fast-forward in either the record or playback positions. Both the fast-forward and rewind speeds are considerably slower than on most recorders, which is a function of motor size. The motor is relatively small and light to maintain portability. The machine was designed to be carried on a shoulder strap, and it weighs 15 lb with batteries installed and tape in place. I carried a Nagra daily on assignments for many years, and I attest to the fact that the 15 lb grew heavier each year.

The front panel of the 4.2, from left to right, has a manual-automatic switch that determines whether the recorder will have manual gain control on both mike inputs

automatic gain control on mike 1, or automatic gain control on mikes 1 and 2. Below this control is a low-impedance earphone jack (50 ohm) with a spring-back waterproof cover, and a recessed, screwdriver-operated earphone-level control.

Next is the Nagra 4.2's meter. The machine normally comes with a PPM reading modulometer, or it may be ordered with a VU meter. In either case, the meter face has several scales. The various scale readings are a function of the meter selector switch, which is next and to the upper right of the meter. Selectable are audio level in the top position, then battery reserve, volts per cell, compression, in the automatic level modes, motor current, and record bias current. These readings can diagnose quite a bit about the operation of the machine for its operator.

Directly below the meter selector is the L, or filter selector. This switch chooses a flat recording response or one of four positions of low-frequency roll-off that aid materially in combatting ambient noise while recording or one of three high-pass filters.

The next three items on the front panel are the mike 1 rotary pot, the line-input, playback output pot, which also serves as the mike 3 input pot when the optional BS preamp accessory is plugged into the Tuchel DIN connector on the left side of the machine, and then the mike 2 rotary pot.

Following the input controls are three recessed toggle switches. The upper is an A/B monitor selector that provides both the earphones and line out with either the input signal or the tape playback signal. The middle toggle provides the same A/B function for the meter, and the lower toggle provides switching of input power from either batteries or AC. If batteries are in place and AC power fails, the 4.2 will automatically switch over to batteries.

The rotary main mode function selector, which is next, may be set to choose one of three positions above the horizontal stop: all-power-off position, or one of the two positions below it, which are playback, where tape signal is fed to the output banana-jack type of terminals on the right side of the machine, and loudspeaker, where signal is fed to a small internal speaker on the machine's right side.

The first detent *above* the horizontal is the test position, in which all the amplifiers are energized, except for record functions, and the puller is inoperative. This position is used for listening, taking levels, or feeding a line without recording. The second position is the standard record mode with the recorder's limiter functioning, and the third position is record mode without the limiter.

The last two items on the right side of the front panel, past the mode function selector, are Sasse indicators, mechanical devices whose faces are normally black but become white crosses on a black background when indicating something. The upper Sasse indicates pilot tone failure (from a camera) and the lower indicator, speed or power failure when the 4.2 is operating in conjunction with a motion picture camera in synchronized double-system moviemaking.

The left side of the Nagra 4.2 has two XLR-3 mike input connectors, line input banana-jack type of terminals, and Tuchel DIN input accessory connectors.

The right side has the grille of the internal monitor speaker, the ATN–external power DIN connector, the line output banana-jack terminals, machine-grounding terminal, and DIN connectors for a whole series of pilot tone accessories used in conjunction with a motion picture camera.

The head assembly on the top panel of the 4.2 has four heads. The first by itself on the left is the erase head. The three heads clustered together are the record, pilot tone, and playback heads.

Several types of optional-cost preamps

are available for the 4.2, some to match specific Sennheiser or Neumann mikes, preamps that provide for simplex or phantom powering of condenser mikes, and universal, general-purpose preamps.

The battery supply of twelve 1.5 V D cells will last $8\frac{1}{2}$ hours in continuous use or 18 hours if used for 2 of every 24 hours.

The Nagra 4.2 measures $13\frac{1}{8}$ by $9\frac{1}{2}$ by $4\frac{1}{2}$ inches.

The performance obtained by recording with a nominal level of 0 VU is equal to 320 nW/m at a maximum peak level of +4 dB.

The frequency response at −20 dB is 15 ips, 30 Hz to 20 KHz; 7.5 ips, 30 Hz to 15 KHz; 3.75 ips, 30 Hz to 8 KHz. Speed stability at all three speeds is ± 0.1%.

The Nagra IV-S stereo recorder (Figure 15–5) is in most ways very similar to its monaural counterpart, the 4.2, with some of its controls and switches mounted differently, but in addition to its two track stereo input and output it has one very unique feature. This feature is only available with the PPM modulometer: the meter has two independently operated needle indicators, one red, one black within the one meter, instead of the two meters one would expect to find on a stereo recorder. When the operator's eyes get accustomed to this feature, it becomes much easier to read the stereo signal pair than to split one's vision between two meters. Another major feature of the Nagra IV-S is that it employs SMPTE-EBU time-code, recorded longitudinally on a center track of the audio tape. This permits time-code editing as well as synchronizing playbacks.

Sony Recorder

The Sony PCM 2000 (Figure 15–6) is providing competition for the Nagra recorders. A portable, compact, lightweight, digital R-DAT format recorder, it is $8\frac{3}{8}$ by 3 by $10\frac{3}{8}$ inches and weighs 8 lb

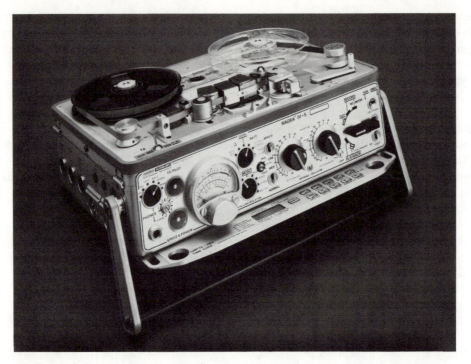

Figure 15–5 The Nagra IV-S stereo recorder-reproducer. (Courtesy Nagra Magnetic Recorders, Inc.)

13 oz with its rechargeable battery installed. The battery has an operating life before recharge of approximately 2 hours. The PCM 2000 uses the Sony digital cassettes, which are smaller than compact cassettes, and come in 60, 90, and 120 minute sizes.

The PCM 2000 has four direct-drive motors and is controlled by a digital servocircuit for stable tape travel even under adverse outdoors conditions. The PCM 2000 can suffer, on the other hand, from the effects of high external moisture, dampness, or dew, which will hinder the machine from immediately operating. Digital input and output are provided for easy connection to studio digital equipment, and A/D and D/A converters are provided as well for connection with analog equipment. Recordings can be made at one of four sampling frequencies, and the 2000 uses one of the DAT format's

auxilliary longitudinal tracks for recording and playing back SMPTE/EBU timecode.

The following recorder functions can be displayed on the front panel multifunction LCD (liquid crystal display): sampling frequency, tape counter—elapsed time or tape time remaining, start, record or erase identification, transport mode, search mode, caution warning, servomode, emphasis indication, DEW (excessive humidity warning), and low battery warning. Additionally, a level meter section of the display has a 28 segment record-level indicator, an over-level indicator, a battery voltage indicator, and an RF output-level indicator. To the right of the LCD display are record and pause buttons. The top of the PCM 2000 has the pop-up cassette door, and to its right and recessed, are the play, stop, rewind, fast-forward, and search controls. The

Figure 15–6   The Sony PCM-2000 digital recorder-reproducer. (Courtesy Sony Corporation.)

right side has two XLR-3 analog inputs, analog input Z switch, input pad switch, mike phantom power switch, mike low-cut switch, and left and right line out jacks. There are also a phones-out jack and phones volume control and digital in and out XLR-3s. The left side has an external power connector XLR, and coaxial BNC connectors for external synchronization and time-code in and out.

Cassette Recorder-Reproducers

The tendency is strong to disparage cassette recorders in their broadcast application. The feeling is strong on two counts: the cassette recorder eliminates the remote operator's job, and the reporter can do the job alone, but also because the device was originally conceived as a cheap consumer product with minimal frequency response. Early models of the cassette recorder stressed the "turn it on and forget it" aspect and were able to make few claims to quality reproduction.

Today's consumer-level quality cassette recorders do a serviceable job of recording and reproducing as long as the recordable material is within the 100 Hz to 9 KHz range. The interview—the human talking voice—is certainly in that frequency range. The major drawback to cassette recording is that the tape format precludes cut-editing without dubbing to single-track $\frac{1}{4}$ inch tape first. Very few reporters can do an interview without occasionally rephrasing a question or cutting the length of a respondent's answer.

The bottom line in this discussion is that cassettes are used extensively and often well by many reporters at many broadcast facilities.

Sony Cassette Recorder-Reproducers

The Pro II (Figure 15–7) is Sony's professional version of the cassette recorder. It is a four-track two-channel stereo machine. It has Dolby noise reduction, and its frequency response, with a chromium oxide tape cassette, is 40 Hz to 16 KHz, ± 3 db, with the Dolby circuit off. Its THD is 0.9%, and its wow and flutter is rated as 0.06% under the same circumstances. The Pro II has two XLR-3 mike level inputs and two line-out phono jacks on its right side. It has a stereo earphone

Figure 15–7 The Sony TC-D5 Pro II cassette recorder-reproducer. (Courtesy Sony Corporation.)

jack and a built-in speaker. It can be powered by 2 D alkaline batteries, with about a 4.5 hour operating life, or by a car battery or AC adaptor. It is approximately 3 lb 12 oz and $9\frac{5}{8} \times 2 \times 6\frac{5}{8}$ inches.

The TCM 5000 (Figure 15–8) is a two-track single-channel monaural recorder with a frequency response of from 90 to 9000 Hz and a three-head recording-monitor system. There is a counterinertial flywheel system for tape transport stability when tape is in motion. Recording can be performed with manual or automatic control. The concentric level controls are arranged so that the outer ring controls mike 1 and the inner either mike 2 or playback level. The VU meter also reads battery supply. There is an A/B—source or tape, switch, and a variable-speed control for playback that provides +40% or −20% control. The TCM 5000 weighs 3 lb 3 oz with its 4 C cells installed, and its dimensions are $2\frac{1}{4}$ high $\times$ $10\frac{1}{8}$ wide $\times$ $5\frac{3}{4}$ inches deep. Twenty hours of recording are available from one set of four alkaline batteries, or the recorder can operate on an AC or DC (car) power adaptor.

Figure 15–8 The Sony TCM-5000EV cassette recorder-reproducer. (Courtesy Sony Corporation.)

REMOTE BROADCAST-RECORDING PROCEDURES

When a decision has been made to do a remote, and if it is to be a line feed, an order is placed with the local telephone company for a line installation. Very explicit instructions are given to the telephone company about what is needed—unequalized or equalized line and to 5 or 15 KHz, coupler or no coupler, the amount of time that the line is to be up, the start time, and where the line should be terminated at the remote site. Often an operator is dispatched from the station to show the installer where the termination should be made and to do a site survey. In addition to the radio loop, a video loop may be ordered, and a temporary business phone is usually ordered to the same site location to be used for cuing and other production purposes.

The telephone company installer terminates the radio loop as ordered, in bare ends, in a 42A terminal block (Figure 15–9), in one of the modular connector jacks now in use or in a coupler. The installer uses a "quad" cable, four solid wires color-coded red, green, yellow, and black, which she runs from a nearby telephone company mainframe on or near the premises to the customer's equipment. Normally, the red and green wires carry the radio loop and are tagged with the line designation, which reads something like "10PT 1234." The PT indicates that the wire pair is a radio loop. The telephone company tag usually indicates the broadcast station's call letters as well, but it is good practice for the remote operator to bring the 10PT number, which he will get from his traffic department, and they from Telco, to the remote, for comparison purposes, especially if more than one station with lines is covering the event.

If the remote will be complex, or if the site or type of remote is new to the operator, he is often sent to survey it in advance. On site he looks for an AC power source near his proposed equipment setup point, checks the radio line and telephone terminations, estimates in liberal fashion the number and footage of mike cables he will need, and arranges to secure a table and chairs for his use during the remote.

Figure 15–9   42A terminal block. (Courtesy Bell Laboratories.)

PREREMOTE CHECKOUT

Gather all the equipment that has been deemed necessary to do the job in one location. Test every input to the mixer or recorder; try every mike and mike cable; see that the recorder will indeed record. This cannot be stressed too strongly.

When a thorough equipment check is completed, pack everything carefully in a suitcase or other sturdy carrying case.

Each mike should be encased in its protective sleeve, and any other fragile item such as the mixer should be wrapped in plastic foam sheets or protected by crushed newspapers if it does not have a protective case of its own. Make sure that nothing rattles around in the carrying case and that nothing has been left behind.

Check the time accuracy of your watch setting against the control room clock and reset the watch if necessary. Leave the station in plenty of time to allow for traffic delays.

ON-SITE EQUIPMENT SETUP

On the previously obtained table, set up the mixer or tape recorder. Then place the mikes, on their stands, in the approximate vicinity of their final pickup points. Connect mike cables and necessary extension cables to the mikes, tagging each cable with masking tape at the mixer end as to mike placement, and dressing the cables as far away from foot traffic paths as possible. Leave several feet of cable coiled at the base of each floor stand to allow for movement of the mike. Tape down the cables with 3 inch wide gaffers or duct tape wherever there is the possibility of danger from moving feet.

At the mixer, connect the mike cables one at a time, checking each mike before proceeding to the next. With power on, and earphones connected and worn, plug in a mike cable connector to the mixer or recorder, open its pot only, have someone scratch the mike face (for positive mike identification) but not blow into it to avoid damage to the transducer element. Mark the mike function (e.g., piano, lectern) on a piece of masking tape and place the tape on the mixer close to the pot. Close that mike pot and duplicate this procedure until all mikes are hooked up and checked out.

It is not acceptable practice, under any circumstances, to tape your microphone to someone else's mike or mike stand. There are mike clamps that may be used for that purpose, but only with the prior permission of the other mike's owner.

If the mixer is to feed a recorder, connect the output of the mixer to the input of the recorder. Be careful to observe impedance matching. If the mixer has a mike level output (as does the Shure M267), then that output can be fed directly into a recorder's mike input. If the mixer has only a 600 ohm line output, then it must feed the line input of the recorder. Alternately, it can feed a recorder mike input through a line-to-mike matching transformer adapter.

If the mixer or recorder is to feed a dedicated, or equalized, program line, connect the output of the recorder to the line. On remotes that use nonequalized, subscriber-service telephone lines for program transmission back to the studios, the telephone company requires the use of an interface, called a coupler, between the user's equipment and the line unless there is a hybrid coil at the studio end. The coupler protects the feeding and terminal ends of the line from feedback and ensures proper impedance matching. The QKT voice coupler (Western Electric type 30) may be supplied by the telephone company at additional cost, or the operator might carry, as part of his equipment, a coupler such as the Elgin 30A or the Comrex TC-1.

Grounding, or providing a common ground for all equipment, is often necessary in remote work to eliminate high hum levels. Hum noise, often caused by "ground loops" or by a difference in AC ground potential between items of equipment, is eliminated by providing a common ground between equipment. The grounding procedure is quite simple. A single piece of copper wire is connected to the ground lug or, if none exists, to a screw on the equipment chassis of each piece of equipment (mixer, recorder) and then to an electrical ground such as a cold water pipe or to the cover-plate screw located between any two AC receptacles, which holds the cover plate over the receptacles. Grounding is usually unnecessary if all equipment is battery powered.

The rationale, explained earlier, behind connecting mikes first to the mixer, checking their operation one by one, and only then connecting the mixer to the line or to the recorder is that any operational problem that might arise can be more quickly traced to its origin with a minimum of interconnected equipment.

Assured, then, that all equipment to be used on the job is connected and working satisfactorily, proceed to set levels with the station. Call on the business phone and feed a test tone at zero level down the radio line. The source of tone may be internal to the mixer or may be from an external tone oscillator fed into a mixer mike input. A fellow operator at the station, on receiving tone on the line, adjusts the remote line pot for a corresponding zero level reading on his console VU meter.

If the event is to be recorded on site only, then 30 seconds of zero-level test tone should be recorded at the beginning of the tape as a reference level.

When no tone source is available preceding a line feed, then we resort to an ancient level-setting method known as "woofing peaks," which is performed as follows: Sit in front of the mixer observing the VU meter with only one input pot open and its mike directly in front of you. Call your colleague at the station, and when he indicates that he can hear you on the radio line, say "woooof" into the mike and immediately thereafter call out the highest reading or peak for that woof on the VU meter. For instance, "woooof,

plus two; woooof, minus one; woooof, zero." The operator at the station adjusts the line input pot to approximately match the peaks called in by the remote operator.

After checking levels with the station, load tape on the recorder, take levels on the participants if feasible, refine the mike setup, and acquire a printed program of the event or advance texts of speeches if they are available. The printed program assists in knowing what not to record or feed. The advance text of a speech (if the text is followed by the speaker) aids a news editor who wishes to take short news clips out of the taped speech.

If the event is to be recorded only, roll tape at the beginning of the event or at the beginning of the part of the event requested by the station. If the event is to be broadcast live and has been coordinated with the station to begin at a given time, call the station 5 minutes ahead of the start time to coordinate start cues. Open a mike or mikes to feed room noise or audience chatter down the line. The operator at the studio should have the remote line open in audition on her console. At the cue to start, she brings the remote key to the air position and the program is in progress.

PA SYSTEM PICKUPS

It is generally advisable not to take PA feeds. The occasion will arise on a remote job, however, where it is either not feasible or not permissible to place microphones because of event producers' insistence that a forest of mikes on stage (yours and the PA system's) would obscure the performance from a paying audience. The problem of duplicating mikes on stage may be overcome by arranging to take a feed from the indigenous PA system or to "split" mikes with the PA operator.

Operationally, a PA feed presents no difficulty. Take an output of the PA system—usually at very high level and very low impedance (up to 70 V of power at 4 to 16 ohm impedance) through a cable compatible to both the PA output and your equipment input. Then "bridge" the PA feed through either an external bridging coil to a mixer mike input or through a bridging line input directly to the mixer. Grounding the mixer (or recorder) may be necessary to prevent hum from the PA system. Gain control is exercised through a single input pot, which makes the operator's job easy during the performance. What you get, however, is always of dubious quality. PA systems are typically badly maintained, and, once set up and turned on, they are often left unattended during the performance.

PA mixing is adequate at best for speech only. Aesthetically, a PA feed can be dreadful, particularly if the event is a music program. When music is involved, or when recording theatrical or concert performances, the miking technique for broadcast or recording differs significantly from the technique used for sound reinforcement or public address. Typically, a PA operator only mikes those instruments that need sound reinforcement in that particular environment. He is there, furthermore, to serve the house audience, and a broadcast feed in addition is his very lowest priority. The tympani get no miking in most cases, and that is as it should be for PA work. For recording or broadcast purposes, however, the net effect is of a very bad mike setup, especially if directional PA mikes are used.

To overcome this problem, the broadcast operator may arrange to "split" mikes with the PA. Each mike is fed to a passive device that matches the mike impedance and provides two separate outputs of that mike, one for the PA system and one for the broadcast mixer input. The broadcast operator then separately mikes (if he can)

those instruments omitted by the PA and, at the very least, has control of the orchestral balance with whatever mikes are available to him.

MULT PICKUPS

In major metropolitan cities the network affiliates employ "mults" (multiples) at news conferences and speeches by prominent figures in the news when coverage of those events is expected by several electronic media representatives. A mult (not to be confused with patching mults) is a unity gain amplifier with two or more mike inputs and many (perhaps several dozen or more) mike level outputs. Well-constructed mult boxes are always expensive items, primarily because of the transformers necessary at each output to isolate one from another. Consider the chaos if a short-circuited cable were plugged into a mult output and there was no interoutput isolation. Increasingly as well, mults offer both mike level (-50 to -40 dBv) and line level (-10 dBv to $+4$ dBv) outputs, with both balanced and unbalanced terminations available.

For a typical speech remote, the mult is set up with two mikes in front of the participants, and each covering station takes a mike- or line-level feed from the mult output to a mike or line input to its equipment. This facility eliminates a multiplicity of microphones in front of a speaker with some mikes of necessity being poorly placed for optimum pickup.

One admonition regarding either PA or mult feeds is that the feed will only be as good as the operator and the equipment providing the feed. Another is that union jurisdiction may prohibit taking a feed from a nonunion operator or from an operator who is a member of another union.

A final admonition concerns the use of station logos on microphones at remotes. In some cities there is a blanket prohibition by mutual consent (Washington, DC, for instance) of their use when there are two or more station's mikes at an event.

AURAL QUALITY OF REMOTES

A broadcast or recording studio is designed acoustically to enhance its program output. A remote pickup will, by its very nature, have a different acoustic "flavor," and no attempt should be made to match studio quality. One of the programmatic essences of a remote is that its sound quality is different from studio sound. Background sound that would be intolerable in a studio is an essential part of remote sound. Therefore, do not be overly concerned about airplanes flying over an outdoor pickup, coughing in the audience, or off-mike questions at a press conference as long as you have the best possible mike setup for the job. On outdoor events, however, when there is a strong wind blowing across the mikes,

some wind noise may distort pickup quality, so wind screens (foam plastic covers) are placed over the microphone heads to diminish wind noise distortion.

REVIEW QUESTIONS

1. Describe the essential equipment used on a remote.
2. What is a mixer? What type of inputs does it generally have? What type of inputs does it generally not have?
3. Describe what an operator looks for on a remote site survey.
4. Describe the preremote equipment checkout in detail.
5. Why do we feed test tone or woof peaks?
6. Describe the essential differences between a tape remote and a line feed.
7. Describe the advantage and disadvantage of a PA feed.
8. What is a wind screen?

16

THE EFFECTS OF STUDIO DESIGN ON SOUND QUALITY

The characteristic sound, in terms of brightness or lack thereof, and sometimes referred to as its "coloration," of a studio, concert hall, or auditorium, although based on acoustical engineering design principles, is to a large degree the result of empiricism. Particular building materials and specific furnishings do affect sound by either absorbing or reflecting it, but how much absorption or reflection takes place determines the studio enclosure's sound characteristics.

Factors other than physical design affecting studio sound include the presence of human sound absorbers and the amount of humidity in the air.

We speak of air as an elastic medium. High humidity makes air more dense and less elastic. When the mechanical vibrations of sound in an air medium impinge on the surfaces of another medium—walls, floor, ceilings, vibrations are reflected back into the air if the surface is hard and nonporous, absorbed by the surface if it is soft and porous, or transmitted by the surface if it can be made to vibrate sympathetically with the impinging sound.

A room's characteristic sound can thus be altered measurably by making it more reflective or more absorptive.

STUDIO DESIGN FACTORS

In the design of a room to be used as a studio, the first aural factor to be considered is sound isolation. Sound originating outside of the room should not be allowed to enter through the walls, floor, ceiling, or air-conditioning ducts. Sound originating in the studio should not be permitted to escape. Sound isolation of a studio area is accomplished by soundproofing—the construction of the room using heavy porous wall material (such as cinder block) or covering the walls with soft absorbent fiber material like carpeting; covering the ceilings with Celotex tiles or acoustic plaster; and covering the floors with cork or heavy carpeting. The air-conditioning ductwork may be fitted with baffles that deflect sound until it decays. Sound-absorbing material such as felt covering placed inside the ductwork will absorb sound but over time will also absorb moisture and will become a breeding ground for mold and bacteria, causing respiratory illnesses in the studio's users.

Another studio soundproofing method, more costly and less often used, is to suspend a room within a room on springs or rubber material that will not transmit sound.

The second aural factor to be considered in studio design is the control of sound quality within the room. This control is exercised by controlling the reverberation paths and reverberation time within that room.

Reverberation time is defined as the time, expressed in seconds or fractions of seconds, that elapses between the cessation of a sound and the decay of that sound to one millionth of its original decibel value. A studio constructed for live music, for example, will have a greater designed-in reverberation time than one constructed solely for speech. Materials that control reverberation time, by controlling sound absorption of their surfaces, are the same as those used for soundproofing, plus hard surface material such as masonite paneling for reflection and wood surfaces for sympathetic vibration.

Reverberation path is the route that a sound wave takes from its point of origin, through reflective surfaces, back to its point of origin. Reverberation path may be controlled primarily by the relationships, in terms of angle, of wall to parallel wall and floor to ceiling. The further from parallel that these surfaces are to each other, the longer the reverberation path will be. A room designed as a perfect cube would have the shortest reverberation paths. A room designed with no two surfaces parallel would have a much longer reverberation path. That is why one sees studios with strangely slanted ceilings and walls at odd angles to each other

A secondary and continuously adjustable method of controlling reverberation path is by use of tiltable wall panels for varying wall angles and movable screens in the studio (called *gobos*), similar to the screens used in business offices, which in the studio can be either reflective or absorptive.

A condition of "standing waves" can occur in a room whereby sound waves are reflected back on themselves, causing sound or frequency cancellation at certain points within the room. This condition, if unchecked, would cause very poor quality in that room's recording or broadcast of music; the room's sound-reflecting angles or path and the absorption must be changed.

A studio with long reverberation time and short reverberation paths is termed a live studio; conversely, one with short reverberation time and long paths is called a dead studio.

LIVE-END, DEAD-END STUDIO

The live-end, dead-end studio has the same characteristics as a live performance auditorium, where the stage area is highly reflective to sound and the audience area is relatively unreflective. This dual reflective response can be designed into a studio without an audience area by making one end of the studio virtually dead to sound reflectivity.

Thus, the live end where the music is performed enhances musical brilliance and overtones, while the miking is accomplished in the dead end, where reflectivity would color the response of the pickup.

AUDIO IN THE TELEVISION STUDIO

Sound quality in television production has almost always been considered to be of secondary consideration to picture quality, and this is unfortunate. Television sound only reaches prime importance in its total absence, when there is audio equipment failure.

Television studios are often large, barn-like structures necessitated by the inclusion of sets, scenery, and props to build a visual illusion and bulky camera pedestals and mike booms that must be moved around during programs.

The lighting of the television scene requires extensive overhead gridwork, to which the lights are attached and from which they are aimed. This necessitates that the room have a high ceiling, often two or more floors high.

All this adds to the burden of audio control. Reverberation paths are more complex and very long; the mikes are often necessarily far from the performers; and hollow, echolike, roomy sound quality could be the result without the use of hypercardioid mikes. For these reasons, the internal surfaces of a television studio are designed to be as sound absorbing, or dead, as possible.

The reverberation time of a studio may be calculated by employing the Eyring formula:

$$T = \frac{0.05\ V}{-S\ \log e\ (1-a)}$$

where: T = reverberation time in seconds

V = volume of interior in cubic feet

a = average absorption per sq. ft. of surface area

S = total surface

log e = natural or naperian log (not log 10)

Using the reverberation time arrived at by the formula above, it is possible to determine the optimum reverberation time for a given broadcast studio by applying the results of the formula to the Morris-Nixon curves (Figure 16–1).

Figure 16–1   Morris-Nixon curves.

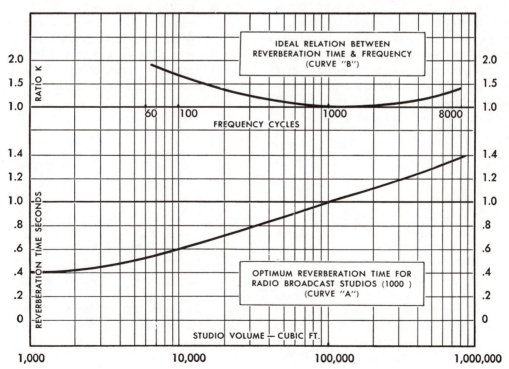

The Morris-Nixon curves plot reverberation time, in seconds, against studio volume, in cubic feet. First calculate the reverberation time at 100, 250, 500, 1000, 2000, and 4000 Hz. To determine the optimum reverberation time at each of these frequencies, first obtain the volume of the studio in cubic feet. From curve A of Figure 16–1, the desired reverberation in seconds for 1000 Hz can be obtained if the user knows the volume of the studio. The reverberation time at the other frequencies is expressed as a ratio to the 1000 Hz reverberation, as shown on curve B. Multiply the number of seconds required for 1000 Hz by each of these factors to determine the reverberation for the other frequencies.

CONTROL ROOM ERGONOMICS

Control room and studio design in a large measure affect control room ergonomics. Ergonomics can be thought of as the manner in which the human body interfaces with its work environment. That environment in the audio operation framework should encompass seating comfort, console operation comfort, that is, everything within hands' reach, pleasant surroundings such as brightly painted walls and good but not glaring lighting, clean work surfaces and floors, and lots of fresh air. It includes as well the listening environment discussed earlier.

It has been traditional in broadcasting for the audio operator to have the most comfortable chair in the control room. He is often in it for hours. An ergonomically well-designed console, if it is a large one, will curve in semihorseshoe fashion around the operator. Turntables and tape playbacks will be placed so he can reach them without straining. In short, a comfortable operator, in bright surroundings, will assuredly perform at his best.

REVIEW QUESTIONS

1. How may a room be soundproofed? Describe the materials used.
2. What is a floating studio?
3. Describe sound reflection, sound transmission, and sound absorption.
4. Describe sound reverberation. What are its two components?
5. What materials are used to control reverberation? How are they used?
6. What is a live-end, dead-end studio?
7. What factors are involved in the aural quality of a television studio?

17

TRADE UNIONISM IN THE BROADCASTING AND RECORDING INDUSTRIES

Most of what you read in this text is based in fact or operational certainty. This chapter, however, is mainly opinion—my opinion. I am overwhelmingly for trade unions in the broadcast and recording industries because I have seen over the years how salaries and working conditions can be affected by both the nonunion and the union-organized broadcast operation.

Let us set up this discussion by stating two divergent points of view that ask questions having no clear-cut answers.

Point one: In a highly competitive field such as broadcasting and recording, where high-paying operator-technician jobs are hard to get and hold, any attempt by an individual employee to deal effectively with his employer on his own concerning pay, working conditions, work shifts, vacations, or any other personnel matters is sheer foolhardiness and will, if the employee persists, ultimately cause him to be fired, no matter how skilled he is.

Point two: Although politically we are all created equally, some of us after a while are much better at our jobs than are oth-

ers. Trade unionism by definition must see and treat all of us as equals within the framework of seniority on the job. That is, the operator who has been on the job the longest has the greatest seniority and, in the case of job lay-off, must be let go last. Therefore, the unions must and do insist that employers treat us equally within the same seniority framework.

The question without a ready answer, then, is, How do we as individuals who are confident of our own superior worth on the job assert that claim within the structure of a union-organized job?*

Trade unionism in our industries began for all the traditional reasons: salaries were lower than they should have been and conditions in the work arena were bad— long hours of unrelieved tedium, unsanitary and electrically hazardous work sites.

The broadcast and recording industries

---

*By way of credentials for this discussion, I served two terms as president of my local union of broadcast technicians and was later a member of broadcast management for some years until retirement.

are relatively small industries in terms of the number of employees at a given place of employment. The operator-technician's entry into the trade union movement was initially through unions that were, in two out of the three major cases, created by other employees in related industries. We are a fragmented industrial group, organized primarily in only the larger United States and Canadian cities, into one of the three major unions for the most part: the International Brotherhood of Electrical Workers (IBEW), the International Alliance of Theater Stage Employees (IATSE), and The National Association of Broadcast Employees and Technicians (NABET), all affiliated with the AFL/CIO. A much smaller number of operator-technicians are organized in smaller and often nonaffiliated unions, such as the National Federation of Federal Employees (NFFE), a union of federal government employees.

Trade unions are said to be organized either vertically or horizontally. Vertically organized unions embrace only those members of a particular craft, while horizontal unions enroll members of any craft or skill within a particular industry or company.

MAJOR UNIONS

The IBEW is an old-line vertical craft organization of primarily electricians' locals, with some few local unions of broadcast engineers and technicians.

The IATSE is a vertical craft union composed mainly of stagehands, stage electricians, motion picture camera operators, and projectionists. IATSE represents operator-technicians at some television stations.

NABET was, it has been alleged, formed as a company union by NBC and ABC. Originally it was only for technicians, but it became an independent union organ-

ized horizontally. NABET now represents other broadcast workers (announcers, secretaries) as well as technicians.

Trade unions are organized as locals, each with an identifying number, of the parent organization, and we will discuss the union local as we proceed.

The newcomer to the broadcast-recording industries surveys these three giant trade unions and asks, especially since they are all AFL-CIO affiliates, why don't they get together into one large strong labor union? The response is very complex, but for this discussion, suffice it to say that each union has differing aspirations and goals for its membership and that each, as in any human endeavor, is steered by a group of people with a natural degree of self-interest at stake. Thus they remain separate and apart. They rarely cooperate with each other against what they perceive to be the common foe. They often, in fact, cross each other's picket lines. Large national trade unions are not themselves homogeneous. They are composed of remarkably independent and individualistic local unions that are, by that very individualism, the backbone of the trade union movement. These local unions can, and often do, act independently of their parent groups and indeed occasionally secede from one parent group and join another.

The newcomer then considers, "Must I join a union? I have been told that dues and assessments are expensive. What do I get out of joining? Don't I surrender some of my rights to self-determination if I join?" And the newcomer has many other similar questions if she or her family has no trade union background.

Know then, with absolute certainty, that if you want to be a technician-operator in the broadcast-recording industries, in any of the larger cities and at a major company, that you not only should but in most cases must join the union that is in place at that company. The contract that was agreed on between the company's management and the union representing its employees generally stipulates that new employees must apply within a short time after being hired to join the union. Note here that the union need not accept you as a member, for whatever reasons they may deem fitting. Perhaps they have discovered that you were a "scab" during a labor dispute. You must, however, make application to join to maintain your job. Indeed, in some cases you must already be a union member before the employer can hire you.

The dues and assessments that you pay as a member are indeed high, but they buy for you and your family a large measure of security and peace of mind over the period of your working life. The rights that you surrender to the group are the same or similar to those rights that as individuals we surrender to government, and they are for the "greater good."

As to one's assertion of individual superiority of skills and talents mentioned earlier, suffice it to repeat the adage that "cream rises to the top," and a fair amount of superior workers are recognized and rewarded in every field of endeavor. Of course, some high-level people are incompetent, as in every field.

Supposing, then, that you could work at a job where everyone else working at the same or a similar job was a union member, but you need not be. Would it be fair for their dues, their organizational efforts, their fights for pay increases and added work benefits—in short, their union—to carry you without your participation? Would you expect to have many friends in the workplace? Moreover, would it be in your best interests for a union to which you did not belong to represent you before management without your having a voice about what that representation should be? On balance, in my judgment, the scale tilts heavily in favor of union membership.

And what of management? One would at first think that it would bitterly resent trade unions. In the early days, it did indeed and fought hard against their organizing. Enlightened management these days, however, uses the trade unions as yet another layer of supervision of its employees, and skillful management makes ample use of that facility. Trade unions, for example, police many of the areas concerning their members that management finds difficult. This includes pressure for advanced technical training, and it includes keeping employees with explicit personal problems (e.g., alcoholism, drugs) that affect the job "in line." The local unions engage in these activities, again, for the greater good, to keep the level of job professionalism high, in the hope that, in turn, salaries will remain high.

The same contract that spells out the employee's rights, often in minute detail, also enumerates management's rights vis-a-vis the employee as well. Skillful and light-handed use by management of contractual prerogatives gives the company a cost-free layer of supervision, which in many areas would be out of reach.

LOCAL UNIONS

The local union, a chartered entity (numerically designated) of the parent national or international union, is perceived to be three tangential spheres: the "body," which is the membership; the executive board, an elected group of officers of the membership; and the business agent, if the local is sufficiently large or monetarily endowed enough to afford one. The business agent is normally a full-time appointed official of the local. He works solely for that local and is often chosen from the local's ranks for his acumen in the area of collective bargaining and for his overall business ability. If he is a member of the local, then on his appointment as business agent he receives an extended leave of absence from his employer, without pay. The business agent's salary is often computed and paid by the local as a fixed percentage figure above the highest member salary in the group. Thus, it is to his personal advantage as well as the members' for him to negotiate hard for higher wages.

Negotiation for wages, working conditions, and jurisdiction (who may operate what and where) take place normally every 3 years and results in a document known as an agreement or contract. The union contract, between the parties, spells out in minute detail wages and the timing of wage increases, benefits (e.g., hospitalization and medical insurance), jurisdiction, and other items of importance.

Local unions without a business agent rely on their elected officials to perform the business agent task. The business agent job is a constant balance between the adversary role played with one or another layer of management over grievances, either real or imagined; jurisdictional disputes with management or with other unions; and the unending disputes with individual members and groups of members over whether enough of the business agent's time, energy, and strength has gone into fighting for specific causes and problems.

The local's executive board consists of the elected officers of the local. These officers are the president, one or more vice-presidents, a secretary, a recording secretary who keeps the minutes (records) of the meetings of the local, a treasurer, and a group of shop stewards. The board meets monthly or more often if required by the local's needs. It formulates the lo-

cal's policies. The shop steward is the union officer closest to the members, and it is to him that individual grievances and problems are first brought. It is also from the shop steward that the member gets first-hand advice on union or union-management matters.

The local union meets as a body, normally monthly, to ratify the work of its elected officials and approve or disapprove every action contemplated by those officials, most particularly the expenditure of union funds. The local union can be an expression of democracy at its highest level of perfection, when it works. It sometimes does not work because people are not perfect or because members are apathetic, as expressed by nonattendance at union meetings. Nonattendance leads to the certainty that small groups who do attend may run the local union from the floor of the meeting. These groups will make union policy that, although binding on the entire membership, may only be to their advantage and possibly to the disadvantage of the rest of the membership.

The conclusion, then, is my admonition that when you join a union, be an active member and attend your meetings, if only to protect your enlightened self-interest.

REVIEW QUESTIONS

1. Name the three major trade union organizations in the broadcast-recording industries.
2. What is a local? Its business agent?
3. Discuss the advantages and disadvantages of trade unionism in the broadcast-recording industries.

18

THE AUDIO OPERATOR
AND THE FCC

The FCC is the Federal Communications Commission, a federal regulatory agency. It was created by the Communications Act of 1934, and its purpose is to regulate interstate and foreign communication, originating in the United States, transmitted by radio waves or wire.

One of the primary functions of the commission has been to license broadcast stations and the technical personnel who operate the frequency-controlling equipment of those stations. Every audio operator at a station is therefore directly or indirectly responsible to the commission.

Licensing is necessary because of the program channels that are used for broadcasting in the United States, whether for AM or FM radio or for broadcast television. These channels belong, by law, not to the broadcasters who use them but are owned in common by the people of the United States.

Licenses are issued for finite periods of time: for seven years to radio stations and for five years to television stations. A license can be taken away from a broadcast station for cause, during the term of the license, without regard for the investment the licensee has made in physical plant and equipment. Licenses may also be not renewed on expiration, for cause.

Licenses are issued "in the public interest, convenience and necessity," and they are normally renewed on expiration, but renewal may be challenged by interested third parties or by the commission's examiners.

One of the general powers given to the commission has been the authority to prescribe the qualifications for the operators of a station's frequency-controlling equipment, specifically its transmitters, to classify the duties that those operators may perform, and to limit the licensing of operators to U.S. citizens, except for special cases.

The commission has had stringent technical knowledge requirements for operators, for generations, placing the frequency-controlling duties at a station for most of that time in the hands of an operator holding a "first phone ticket," a first-class radiotelephone operator's license. There were additionally second and third class licenses for operators with lesser qualifications whose jobs required less knowledge.

An operator license was issued on completion of a day-long written examination that included groups of questions called elements, where the applicant displayed knowledge of basic communications law, basic operating practice, and basic broadcasting and technical proficiency. The number of technical proficiency elements increased as the class of license applied for increased.

This was a sensible system, in my opinion, requiring more intensive technical knowledge as the class of license became higher. It also meant that station managers had to pay higher salaries to operators with that additional knowledge. And as transmitters became more automated, now needing little in the way of routine adjustment, the pressure by the broadcasting business to deregulate operator classification intensified to where in 1983 the commission issued a general radiotelephone license to operators, which combined the authorization formerly granted under the second- and first-class licenses.

In November 1985 the commission issued an order declaring that general radiotelephone operator licenses would no longer be valid or necessary for broadcast purposes effective Jan. 1, 1986. Operators who work with transmitters since then may obtain a restricted radiotelephone operator permit simply for the asking. These permits are granted without examination and convey all the broadcast authority formerly conferred by the general-class license.

For technician-operators who valued the respect conferred on them by the formerly meaningful licensing, two organizations have been formed to "certify" them: the Society of Broadcast Engineers (SBE), Indianapolis, IN, and The National Association of Radio and Telecommunications Engineers, Inc., Salem, OR.

An operator's restricted radiotelephone operator's permit must be posted in the room where the transmitter and its license reside, and the operator must abide by the commission's rules governing broadcast stations. An operator who violates the rules may be served with a written notice calling the violation to his attention and requiring a written response to the commission, which may suspend the license of anyone who has done the following:

1. Violated any provision of any act, treaty, or convention binding on the United States that the commission is authorized to administer, or any regulation made by the commission under any such act, treaty, or convention.
2. Failed to carry out a lawful order of his employer.
3. Willfully damaged or permitted broadcast equipment or installations to be damaged.
4. Transmitted superfluous radio communications or signals or communications containing profane or obscene words, language, or meaning.
5. Knowingly transmitted false or deceptive communications.
6. Knowingly transmitted a call signal or call letter that has not been assigned by proper authority to the station he is operating.
7. Willfully or maliciously interfered with any other radio communications or signals.
8. Obtained or attempted to obtain, or assisted another to obtain, an operator's license by fraudulent means.

An order of license suspension would be given in writing and would take effect 15 days after it had been received. During those 15 days the operator could apply to the commission for a hearing on the order of suspension. After a hearing, the commission would affirm the order of suspension, effective immediately, modify the order of suspension, or cancel the order.

There are additionally penal provisions to the Communications Act of 1934 and its revisions. Any person who knowingly and willfully violates any of its provisions shall, on conviction of a first offense, be punished by a fine of not more than

$10,000 or by imprisonment for not more than 1 year. Violation of any rule, regulation, restriction, or condition made or imposed by the commission may on conviction be punishable by a fine of not more than $500 for every day during which such offense occurs.

FCC Inspectors have the authority to inspect all broadcast installations associated with stations required to be licensed, at any reasonable hour. If a station is transmitting, it is considered to be a reasonable hour, and this includes nights and weekends. An inspection determines whether station operation conforms to FCC rules and regulations.

Radio broadcast stations are located in one of two frequency bands: AM stations from 535 to 1605 KHz, with stations required to be 10 KHz apart, and FM stations from 88 to 108 MHz, with 0.2 MHz separation required between stations. Every broadcast station is assigned an operating frequency within its band, a maximum operating power in watts or kilowatts, and a set of "call letters."

Every broadcast station is required to identify itself with those call letters at the beginning (sign-on) and ending (sign-off) of each period of operation and hourly, as close to the hour as feasible, at a natural break in the program. The official station identification to be used at those times shall consist of the station's call letters, followed by the name of the community or communities that it serves and specified in its license as the station's location. The name of the licensee may be inserted between the call letters and the station location. No other insertion is permitted. Other remarks, themes, or anthems may be added either before or after the required identification.

REVIEW QUESTIONS

1. State a primary function of the FCC.
2. Why is the licensing of broadcast stations necessary?
3. For how many years is a radio station licensed? A television station?
4. Explain "public interest, convenience and necessity."
5. How can an operator's license be obtained? Can a Soviet citizen get one?
6. For what reasons can an operator's license be suspended? What can she do about it?
7. What are the penal provisions of the Communications Act of 1934?
8. When can an FCC inspector inspect a broadcast station?
9. How often must a broadcast station identify itself? Describe how.

19

OTHER MEMBERS OF THE BROADCAST TEAM

This text has highlighted the job of the audio control operator almost to the exclusion of everyone else employed at a broadcast station. Many other members of the broadcast team have jobs equal in importance to, or more important than, that of the audio operator. In fact, at most small broadcast stations there is an overlapping of job functions. The inference to be drawn is that once audio control has been mastered, the operator who wishes continued employment at ever-increasing skill levels should continue to grow in his chosen field by mastering other facets of that field.

Automation of equipment in recent years has reduced the complexity of audio control. One national leader of a broadcast union has predicted that within the foreseeable future, the only technical operator at small stations will be the occasional maintenance technician, and this leader has strongly suggested that the youthful audio operator broaden his job horizons in broadcasting. I concur with that finding, and so we will look briefly here at the jobs of the other members of the broadcast team.

GENERAL MANAGER

Every broadcast station, regardless of size, has a general manager, whether by that title or some other. At small stations he may perform other duties as well as those of managing the station, and at large stations he may have a staff of assistants. His primary job is to formulate the overall policies by which the station is operated and to provide leadership to the supervisory people of the various departments, enabling them to carry out his policies.

He coordinates the activities of the various departments of the station to ensure that they function in cooperation and harmony. In short, he is the boss.

COMMERCIAL MANAGER

Sometimes known as the sales manager, the commercial manager coordinates the activities of the broadcast-time salespeople employed by the station.

A broadcast station must earn money to stay in the business (unless it is a noncommercial, educational, or public broadcast station), and the commercial manager's job is to ensure that a maximum portion of the station's airtime is sponsored. Sponsorship of programs and commercial messages are how the station pays for salaries and other expenses. A sponsor is a business person, or an advertising agency working for a business, who pays to have messages concerning that business or its products inserted into, or made a part of, a radio or a television program.

At a small station the commercial manager may be the entire sales staff, and at a large station he might have several broadcast-time salespeople to supervise.

PROGRAM DIRECTOR

The program director's job is to schedule the shows that emanate from the station. He determines that the right type of program is on the air at the right time of day, in accordance with audience desires. In carrying out this function, he hires the announcers and other talent used at the broadcast station.

His jurisdiction includes the news department, the production-direction department, the script or continuity department, the music library, the traffic department, and, of course, the announcers and other talent. The program director may supervise all these departments and individuals at a small station or may delegate this authority to department heads at a large radio or television station.

The program director schedules the work shifts of the staff announcers or delegates that authority to a chief announcer. The other talent and the musicians who may be employed full or part time are also scheduled by the program director.

To understand better the complexity of a program director's job, we will examine a few of the departments that she oversees:

News Department

All the international, national, local, and sports news is compiled and edited by the news department and tailored into newscasts that fit the station's broadcast for-

mat. The station receives all but its local news by subscription to the facilities of press wire services or from affiliation with a network. Wire services, which include AP (Associated Press) and UPI (United Press International) in this country, and Reuters in Europe, supply the station subscriber with teletype printers. From the teleprinters emanates a continuous stream of news copy, feature material, and sports round-ups as well as complete newscasts of varying length and up-to-the-minute timeliness. Networks supply newscasts to local stations at predetermined times of the day or night. Local news is acquired by the station's news reporters and correspondents.

Production-Direction Department

This is the department with which the audio operator will have the closest contact. At small stations, the operator-announcer will be a member of the production-direction department, since he will produce and direct his own shows.

At the large broadcast stations, where the bigger or more complex programs are conceived and put together, production and direction each becomes a highly specialized skill. There, production entails assembling a program from the purchase of a script, to the auditioning and hiring of talent, to the copyright and performance clearances required, to the final budgeting and rehearsal needed. At this point, the director enters the picture. He rehearses the cast, adjusts the program pace to fit the allotted time, edits and interprets the script material, and directs the program in its performance.

Continuity Department

The members of the continuity department assemble or write all the script material used. They write or check all the commercial copy submitted by the sponsors. In addition, they keep a file of all continuity and commercial copy that is aired for future reference.

Music Library

The music library buys, or acquires at no cost, records from recording companies. In addition to acquiring popular and classical records, carts, and tapes, it rents specialized music from program library services that furnish mood music, dinner music, transitional music for dramatic and documentary programs, and complete libraries of sound effects.

The music library stores the records, tapes, and transcriptions thus acquired, cross-index files them, and makes them available to members of the production-direction department on request. The library arranges for performance clearances from the music writers' associations, American Society of Composers, Authors and Publishers (ASCAP) and Broadcast Music Incorporated (BMI), which collect royalties for performance.

At small stations, the task of music librarian is usually assigned to a staff announcer, in addition to his air duties.

Traffic Department

The traffic department receives program information from the program director and furnishes the commercial manager with information regarding unsold airtime. It is the station's business department. Traffic begins and terminates all contracts for airtime with the sponsors. This function includes keeping all commercial copy up to date. Traffic receives the commercial copy from the continuity department.

The traffic department is responsible for typing the daily program schedule and, in earlier times, checking that schedule to determine that there were no similar adjacent programs and no adjacent similar spot announcements with different sponsors. Traffic checks the completed program log to see that performance agrees with its billing to the sponsor, as per the contract with the sponsor. They prepare the sworn statements that commercial messages were given at the time paid for and arrange for the announcer's verification signatures. The traffic department also schedules the individual studios for programs, rehearsals, auditions, and recording sessions.

CHIEF ENGINEER

The chief engineer is in charge of all the operators and technicians employed at the station. He is responsible for the care, maintenance, and installation of all the technical equipment owned by the station.

At small radio stations the engineering staff may consist of the chief and two or three transmitter technicians. If the station employs people who can do more than one job, they are under the joint jurisdiction of the program director and the chief engineer. Otherwise, the chief engineer supervises the audio control operators, transmitter workers, and, at television stations, the video control operators and camerapeople as well.

The chief engineer schedules the work shifts of his personnel and makes recommendations for the purchase of new or replacement technical equipment.

TELEVISION OPERATING CREW

On a television program the audio operator is joined by several other operators, both in the control room and studio, to comprise an operating crew. Although we will briefly mention them here, the specific responsibilities of these other operators are more thoroughly examined in books on television production, such as my own *Television Operations Handbook* (Focal Press, 1984). The other operators are as follows:

Camerapeople

Normally from one to four camerapeople perform the videography on a television broadcast, with up to a dozen or more on a major sports broadcast. Each person has the task of picture composition, correct focusing, and following the shot direction of the program's director.

Video Control Operators

There may be one or more video control operators on a program. The video operator's job in a sense parallels that of the audio operator, riding video gain and controlling the shading of the picture at the video control panel. Modern equipment, however, makes that job virtually automatic.

Switcher

Generally the switcher coordinates the television crew and is the crew chief. He switches or dissolves from camera to camera, or wipes from camera to camera, on cue from the program's director.

Boom Operators

Two boom operators are usually employed on an operating crew. They position the boom-operated mikes at the direction of the audio operator in the control room.

Dolly Pushers and Cable Pullers

This job, although mostly physical, requires that cables be layed out in advance so that the are not in the way of the movement of cameras or mike booms. During the program, these workers, move the mike booms and help push the camera pedestals into position.

Floor Manager

The director's aide on the studio floor, he relays the cues from the control room to the talent during the program.

Stage Electricians

The setup and control of lighting in the studio or on location, before and during the program, is their responsibility. They receive their instructions from a lighting director.

Stagehands

They set up and strike (take down) the sets, walls, backdrops, curtains, and props that are used to create the pictorial illusion on a television program.

REVIEW QUESTIONS

1. Describe the duties and responsibilities of the general manager, commercial manager, chief engineer, and program director.
2. Describe the functions of the news department, production-direction department, continuity department, music library, and traffic department.
3. Who are the members of a television operating crew, and what are their duties?

GLOSSARY OF BROADCASTING TERMINOLOGY

This glossary is intended to provide a basic working vocabulary of the terms found in the industry, but it does not purport to be encyclopedic.

Acetate   An instantaneously cut disk; the material on which it is cut.

Acoustic   Pertaining to the sound environment; material for the control of sound on surfaces.

Active circuit   A circuit containing amplification.

Ambient noise   Background sound.

Ampere   The unit of electrical current or electron flow.

Amplifier   An electronic active device that enlarges signal passing through its stages.

Analog audio   Audio signal that is an analogy, or mirror image, of the sound from which it originated.

Anechoic chamber   An enclosure totally devoid of sound reflection.

Antenna   A structure designed to radiate or capture radio frequency energy.

Attenuator   A passive resistive device that presents an insertion loss to a circuit (*see* Pad, Pot).

Audio   Sound signal within the spectrum 20 Hz to 20 KHz.

Audition   To listen to audio without broadcast transmission.

Background sound   A secondary sound source that is audible behind or under the primary source.

Balance   The blend of gain from two or more sources, so that they are in correct proportion, one to the other(s).

Ballistics, meter   The VU meter uses a method of damping that causes the meter to read average audio level. PPM meter ballistics cause the meter to read audio level peaks.

Bias   A higher-than-audio frequency signal applied in addition to the audio to magnetic tape during the recording process to correct for the nonlinearity of the magnetization process.

Bidirectional mike   A mike with two lobes of sound acceptance, one at 0 degrees and the other at 180 degrees.

Blast-in   A sound input of an extremely excessive level.

Blast filter   A foam filter in or on a mike to prevent blast-in.

Board, control   *See* Console.

Boom   A mike stand that projects the mike horizontally, usually above the user. A baby boom is a smaller version.

Booster-amp   An amplifier of intermediate gain, used between a preamplifier and a power or line amplifier.

Bridge   A short music piece used as a sound transition.

Bridging   An input impedance of 10 Kohm or more; a method of connecting electronic devices wherein there is a gain loss but not a frequency-response loss.

Buss   A common point for a group of amplifier outputs; a point to where amplifiers may be switched without disturbing the impedance balance of that commonalty.

Cans   Earphones; headphones.

Capstan   With an associated tangential pinch roller, maintains constant tape speed in a tape-transport mechanism.

Cardioid   The shape of a unidirectional heart-shaped microphone sound acceptance (polar) pattern.

Cartridge, phono   A transducer that derives its modulation from the sides of a record groove.

Cartridge, tape   A continuous closed loop of lubricated tape on a single reel, encased in a plastic container; the record and playback equipment for that tape.

Carrier   A transmitting wave in the radio frequency spectrum that transports, as its sidebands, modulation in the audio frequency spectrum.

Cassette   An encased reel-to-reel tape; the record-playback system for that tape.

CD   A digitally recorded compact disk; the equipment to play back CD recordings.

Chain, mixing   The components, including input, pot, key, or assignment switch, and a preamp of −50 dB of gain that can feed a buss or program-level channel.

Channel, audition   *See* Cue.

Channel, mixing   *See* Chain, mixing.

Channel, program   The input, pot, and +4 to +8 dB amplifier that can feed a program line or the line input to a recorder.

Clearance   Permission obtained from the authors, composers, or publishers to permit use of material for a broadcast.

Clip   To cut off sharply; a short piece of recorded material.

Commercial message   That portion of broadcast time during which the sponsor's product or service is described to the listening or viewing audience.

Compander   A signal-processing device; an acronym for compressor-expander.

Continuity   Scripted material.

Console   An electronic device that controls gain and mixes, balances, and routes audio. Also called control board or mixer.

Control room   A room, usually adjacent to a studio, where the audio control or video functions are performed.

Cross talk   Spillover of audio energy from one circuit to another; spillover of magnetic energy from adjacent tape layers, also known as print-through.

Cue   A signal to start.

Cue sheet   A rundown sheet of cue-ups for the operator of a multiinsert program.

Cue system   The amplification and listening system on which recordings are cued-up or auditioned.

Cue-up   To prepare a recording for playback.

DAT   Digital audio tape. A system of digital audio recording that uses rotary head digital audio tape recorders. Sometimes called R-DAT.

DASH   Digital audio stationary head. A system of digital audio recording using stationary heads.

dB   Decibel, $\frac{1}{10}$ of a Bel.

dBm   db referenced to 1 mW of sinusoidal power in a 600 ohm line.

dBv   dB referenced to 1 V.

Dead   Not live, as in dead mike, dead studio.

Dead pot   To play back on a closed pot.

Decibel   *See* dB.

Degauss   To demagnetize; to disrupt the magnetic alignment of tape particles.

Digital audio   Audio carried through circuitry, or magnetically or optically, in binary mathematical form.

Distortion, audio   Variation from pure signal quality. Major types are intermodulation distortion (IMD), total harmonic distortion (THD), and frequency distortion.

Dub   To make a copy of a recording.

Dolly   A mobile television camera stand; to move a video camera.

Dynamic microphone   A mike using a moving coil and diaphragm transducer.

Echo   Repeated reflection of sound.

Echo chamber   An electronic or acoustic device for creating reverb or sound delay.

Equalizer   An active or passive electronic device that boosts or cuts a frequency or band of frequencies.

Fade   To lower or increase amplifier gain using a pot or vertical slide attenuator.

Fader   The control used to vary gain.

FCC   The Federal Communication Commission, the regulatory agency in the United States that licenses and controls broadcasters and broadcast stations.

Feed   To transmit electronically from one component or circuit to another.

Feedback (or feedback loop)   A closed loop of audio energy in a runaway state, causing an even louder oscillatory howl. It is called howl-round in the United Kingdom.

Feed reel   The tape reel that dispenses the tape.

Filter   A passive device to cut or boost frequencies.

Flat topping   Clipping the tops of audio waves on both sides of the cycle, causing distortion.

Floating mults   Patch panel jacks connected to each other but floating (unconnected to a working circuit) unless in use.

Fluff   To make a speech error.

Flutter   A high-frequency speed irregularity in a playback device.

Frequency   The number of vibrations or cycles in a given time, as cycles per second, in Hertz.

Frequency response   The degree to which audio equipment maintains its designed spectrum coverage without distortion.

Gain   Audio or video energy volume.

Gain control   A fader or pot.

Gain figure   The specific amplification inherent in an amplifier.

Grease pencil   A soft waxy marker used in audio tape-cut editing.

Groove   The depression cut in a record by the recording stylus. The left and right channels of stereo music are cut into the left and right sides of a record groove.

Ground   The potential reference of zero; to connect components together and then to earth for a commonality of potential.

Head gap   The space between the pole pieces of a tape transducer.

Head, tape   The transducer in a tape recorder.

Hertz   Hz, cycles per second.

Impedance   Z, a measurement of alternating current electrical resistance to current flow in a circuit.

Input   That part of a circuit or component into which energy is fed; the front end of a circuit.

Insert   A recorded piece of a program.

Jack   A female connector with a male name.

Jack panel (or bay)   A field of female connectors (*see* Patch panel).

Key   A switch used to connect or mute input components to a buss on a control console.

Key in   To bring in modulation with a key.

Key station   A network-originating station.

KiloHertz   KHz, 1000 Hertz.

Label   The center portion of a record, containing the hole and information such as the artist, the selections, the selection time, and the recording company.

Leader tape Nonmagnetic plastic or paper tape, the same width as recording tape, used to protect the ends, or separate the segments, of recorded tape.

LED Light-emitting diode, a small illuminator used to indicate on-off or peak recording conditions.

Level The amount of audio energy, as read on a VU meter.

Limiter A circuit that protects against signal overload.

Line A two wire circuit.

Linearity A relationship between two quantities, where a change in one quantity is directly proportional to a change in the other.

Live Not recorded; produced in real time.

Log A document relating to programming or technical facts as they occur at a broadcast station, which is or was kept as an FCC requirement.

Loop A wire circuit (*see* Feedback loop); a repetitive closed continuum of tape for reverb purposes.

Loudspeaker A transducer that changes electrical energy into mechanical (sound) energy.

Master control A central program switching point.

Master pot The gain control on the output of a program amplifier.

Matrix A group of interconnected switches used for routing program signal.

Microprocessor An electronic chip (integrated circuit) for processing data.

Mike Microphone; a transducer that changes mechanical (sound) energy into electrical energy.

Miking Using microphones to pick up sound.

Mixer A gain control; a group of gain controls feeding a buss, a consolette used on remotes.

Modulation One frequency, the modulation frequency, superimposed on another, higher frequency, the carrier frequency.

Monaural A single program channel or tape track.

Monitor To listen to program; an amplifier-loudspeaker used for listening.

Mult A unity gain amplifier with multiple outputs (*see* Floating mults).

Multiplexing Using the two sidebands of a modulated carrier wave to carry separate program information.

Mute To lift or disconnect an audio source.

Muting relay A relay operated by a mike key that cuts off studio speakers when a mike is live.

Network A company that feeds program material to a group of broadcast stations.

Noise Random sounds that are not harmonically related and that may occur in any portion of the audio spectrum.

Ohm The unit of electrical resistance or impedance.

Off mike Not within a mike's primary pickup pattern.

Omnidirectional All-directional.

On mike On axis and close to a mike's pickup pattern.

Output The portion of a circuit from which energy flows.

Pad A resistive insertion loss in a circuit.

Panpot A gain control used to fade audio signal from one input or output circuit to another.

Passive circuit A circuit having no amplification.

Patching A system of externally connecting or disconnecting circuits or components using patch cords and patch bays.

Patch panel A panel grouping of equidistant and labeled jacks used for patching electronic circuitry.

Pattern, polar A polar graph of a mike's ability to react to sound waves as they impinge on its transducer either on or off axis.

Peak A rise from the flat portion of a curve; a high reading on a VU meter.

Phase Mikes are in phase when their outputs are additive and out of phase when their outputs are subtractive.

Pickup A phonotransducer; sound received by mikes; a feed from a mult or PA system.

Pink noise White noise fed through a special audio filter that inverts white noise's frequency characteristics and results in a test signal of uniform level (*see* White noise).

PFL Prefader listening.

Platter A phono record; the table on which it is played.

Playback To feed audio from a recording; the device used.

Pot Potentiometer, gain control; alternately called a fader.

PPM Peak program meter, a meter that reads peak program gain.

Preamp A high-gain amplifier.

Processing, audio To feed audio through circuitry that is designed to alter the audio in some manner, usually to remove unwanted portions of the spectrum (noise) or to add boost (gain) to specific frequencies.

Print-through *See* cross talk.

Popping An explosive sound caused by stress in the pronunciation of the letters "B," "P," and "T."

Presence The quality of being on mike at one's correct personal distance from the mike.

Pressing A manufactured copy of a disk recording.

Pressure mike *See* Dynamic microphone.

Program A broadcast presentation with a beginning, middle, and end; program channel (*see* Channel, program).

Puller A tape transport mechanism.

R-DAT *See* DAT.

Receiver A device that demodulates audio or video information from a carrier wave and then enlarges that information so it can be used by a transducer.

Record To make a recording.

Relay An electrically operated switch, either mechanical or transistor operated.

Remix To mix down multichannel recordings to two or four channels.

Remote A broadcast or recording made at a site other than the station's studios.

Response curve A graph indicating a mike's gain in dB versus frequency in Hz (*see* Frequency response).

Residual magnetism The magnetic flux that remains in a substance after the magnetizing force has been removed.

Reverberation Sound reflection; echo.

Ribbon mike *See* Bidirectional mike.

Ride gain To control gain of audio.

Roll-off A diminution or fading out at one end of a response curve.

Segue (pronounced "seg-way") A continuous playing of two or more music records with no break or announcement between them.

Shaped response A design factor in microphone frequency-response curves, resulting in a roll-off or a peak at an end of the curve.

Sibilance The hissing sound made in the overstatement of the letter *S*.

Signal-to-noise ratio A qualitative statement about the performance of an electronic device based on the ratio of the two stated factors. It may also be stated in reverse, as noise-to-signal ratio.

Signal processing *See* Processing, audio.

Sine wave An audio wave whose shape can be expressed as the sine of a linear function of time, space, or both; a tone signal.

Solo Alone; an audio monitor feature wherein one signal out of a group may be isolated for listening without removing that signal from the group.

Sound A disturbance of air particles causing audible vibration of those particles.

Spillover A VU reading of over 0 VU and into the plus (red) portion of the scale.

Sponsor One who pays a fee to advertise a product or service on a broadcast.

Stereo Stereophonic; two complementary channels of recorded or broadcast information.

Studio A room designed to be used for recording or broadcasting.

Stylus The shaped needle with industrial diamond tip that either cuts record groove or transfers the undulations of the groove to the playback transducer.

Talent The person(s) on mike or on camera in broadcasting.

Talkback An intercom system between control room and studio.

Tape A plastic ribbon of specific width, coated with a magnetic oxide and used as a recording medium.

Telco The telephone company.

Track The magnetic signal written on or read from recording tape.

Transcription A record manufactured to be used solely for broadcast.

Transducer A device that changes one form of energy into another.

Turntable A playback machine for records and transcriptions.

Unidirectional One-directional (*see* Cardioid).

Velocity mike *See* Bidirectional mike.

Volt The electrical unit of pressure.

Volume Quantitative amount of sound.

Volume unit One dBm of complex sound waves.

VU meter An instrument that measures average sound volume.

White noise Noise that is uniform over a frequency band; all noise frequencies perceivable to the human ear, heard together.

Wow Low-frequency sound that is off pitch, caused by an off-speed playback.

INDEX